ENERGY AND INFRASTRUCTURE

Volume 2

Beyond the Woodfuel Crisis
People, land and trees in Africa

Full list of titles in the set
ENERGY AND INFRASTRUCTURE

Beyond the Woodfuel Crisis
People, land and trees in Africa

Gerald Leach and Robin Mearns

publishing for a sustainable future

London • Sterling, VA

First published in 1988

This edition first published in 2009 by Earthscan

ISBN 978-1-84407-974-2 (Volume 2)
ISBN 978-1-84407-972-8 (Energy and Infrastructure set)
ISBN 978-1-84407-930-8 (Earthscan Library Collection)

For a full list of publications please contact:

Earthscan
Dunstan House
14a St Cross Street
London EC1N 8XA, UK
Tel: +44 (0)20 7841 1930
Fax: +44 (0)20 7242 1474
Email: earthinfo@earthscan.co.uk
Web: **www.earthscan.co.uk**

22883 Quicksilver Drive, Sterling, VA 20166-2012, USA

Earthscan publishes in association with the International Institute for Environment and Development

A catalogue record for this book is available from the British Library

Library of Congress Cataloging-in-Publication Data has been applied for

Publisher's note
The publisher has made every effort to ensure the quality of this reprint, but points out that some imperfections in the original copies may be apparent.

At Earthscan we strive to minimize our environmental impacts and carbon footprint through reducing waste, recycling and offsetting our CO_2 emissions, including those created through publication of this book. For more details of our environmental policy, see www.earthscan.co.uk.

FSC
Mixed Sources
Product group from well-managed
forests and other controlled sources
Cert no. SGS-COC-2953
www.fsc.org
© 1996 Forest Stewardship Council

This book was printed in the UK by CPI Antony Rowe.
The paper used is FSC certified.

BEYOND THE WOODFUEL CRISIS

BEYOND THE WOODFUEL CRISIS

People, Land and Trees in Africa

by
GERALD LEACH and ROBIN MEARNS

Earthscan Publications Ltd
London

First published 1988 by
Earthscan Publications Limited
3 Endsleigh Street, London WC1H 0DD

Earthscan Publications Ltd
is an editorially independent and
wholly owned subsidiary of
the International Institute
for Environment and Development.

British Library Cataloguing in Publication Data

Leach, Gerald
Beyond the woodfuel crisis: people, land
and trees in Africa
1. Africa. Fuel resources. Wood
I. Title   II. Mearns, Robin
333.75

ISBN 1 85383 031 3

Typeset in 10/12 Times
by DP Photosetting, Aylesbury, Bucks

Contents

Acknowledgements

The research and writing of this book was made possible by the generous support of the Royal Norwegian Ministry of Development Co-operation. We owe special thanks to the Ministry and, in particular, to Per Tobiesen, Erik Whist, Even Sund and Mrs Wenche Gulnes for their patience and encouragement as well as their comments on the report *Bioenergy Issues and Options for Africa* which they commissioned and on which this book is based.

Many other people helped us greatly by sharing their ideas, providing information, commenting on the report or drafts of the book, or assisting us during field trips to Africa. We should like particularly to thank the following for their help in these respects: Martin Adams, Mike Arnold, Ed Barrow, Phil Bradley, Louise Buck, Robert Chambers, Czech Conroy, Marc de Montalembert, Peter Dewees, Musa Enyola, Adriaan Ferf, Willem Floor, Gerald Foley, Carl Åke Gerden, Davison Gumbo, David Hall, Ann Heidenreich, Marilyn Hoskins, Mary Kekhovole, Richard Labelle, Mike McCall, Margaret McCall-Skutsch, Jennifer McCracken, Barry Munslow, Phil O'Keefe, Keith Openshaw, David Pearce, Per Prestgaard, Jules Pretty, Dianne Rocheleau, Ian Scoones, John Soussan, Berry van Gelder, Paul Vedeld, Joseph Wekunda and Paula Williams. These friends and colleagues are not, of course, responsible for the way in which their facts, opinions and experiences have finally emerged in the book.

Gerald Leach and Robin Mearns

International Institute for Environment and Development
3 Endsleigh Street
London WC1H 0DD
England

September 1988

Abbreviations

AFRENA	Agroforestry Research Network for Africa
ASTRA (India)	Centre for the Application of Science and Technology to Rural Areas
Campfire (Zimbabwe)	Communal Areas Management Programme for Indigenous Resources
CARD (Zimbabwe)	Co-Ordinated Agricultural and Rural Development
CARE	Co-Operative for American Relief Everywhere
CEMAT (Guatemala)	Centro de Estudios Mesoamericano di obre Technologia Appropriada
CFSC (Ethiopia)	Community Forestry and Soil Conservation
D & D	diagnosis and design
DANIDA	Danish International Development Agency
DDC	district development committee
DDP (Kenya)	Dairy Development Programme
EEC	European Economic Community
ENDA (Zimbabwe)	Environment and Development in the Third World
ERC (Sudan)	Energy Research Council
ERL (UK)	Environmental Resources Ltd
FAO	Food and Agriculture Organization of the United Nations
FLUP (Niger)	Forestry and Land Use Planning
HADO (Tanzania)	Hifadhi Ardhi Dodoma (conserve soil in Dodoma)
HASHI (Tanzania)	Hifadhi Ardhi Shinyanga (conserve soil in Shinyanga)
ICR	InterChurch Response for the Horn of Africa
ICRAF	International Council for Research in Agroforestry
IIED	International Institute for Environment and Development
IITA	International Institute for Tropical Agriculture
ITK	Indigenous technical knowledge
KDP (Ethiopia)	Konso Development Programme

KEA (Tanzania)	Kondoa Eroded Area
KENGO	Kenya Energy Non-Governmental Organizations
KREDP	Kenya Renewable Energy Development Project
KWDP	Kenya Woodfuel Development Programme
LPG	liquefied petroleum gas
LU	livestock unit
MOERD (Kenya)	Ministry of Energy and Regional Development
NADA (India)	an Indian village from which this NGO is named
NEA (Sudan)	National Energy Assessment
NGO	non-governmental organization
NORAD	Norwegian Agency for International Development
NORAGRIC	Norwegian Centre for International Agricultural Development
ODA (UK)	Overseas Development Administration
ODH	Operation Double Harvest
OECD	Organization for Economic Development and Co-Operation
PA (Ethiopia)	Peasant Association
PADF	Pan-American Development Foundation
PVO	Private voluntary organization
RRA	Rapid Rural Appraisal
SADCC	Southern African Development Co-ordination Conference
SIDA	Swedish International Development Authority
SODEVA (Senegal)	Société de Développement et Vulgarisation Agricole
SPRP (Zambia)	Soil Productivity Research Programme
SREP	Sudan Renewable Energy Project
SSE	Sahel-Sudan-Ethiopia Programme
T & V	Training and visit system
TRDP (Kenya)	Turkana Rural Development Programme
UNDP	United Nations Development Programme
USAID	United States Agency for International Development
VAP (Mali)	Village Agroforestry Project, Koro

Introduction

Much has been written about the plight of developing Africa and the environmental crisis which underlies it. In many parts of the continent food production lags behind population growth, hunger and famine strike with dreadful persistence, soils are degrading, forests and trees are disappearing at unprecedented rates, and poverty deepens in the countryside and cities.

It is not like this everywhere. Nor are prospects for the future as hopeless as these headlines imply. On the contrary, in many places hard work and creative innovations by people, governments and aid agencies are doing remarkable things to put the land into good shape, increase food production, restore soils and a healthy cover of vegetation, and generally enhance livelihoods on a sustainable basis. How best to support and amplify these efforts has become one of the most urgent and challenging tasks of our times.

This book joins the "literature of hope" rather than of despair by presenting these challenges and the opportunities they offer, while recognizing the problems to be overcome. However, it approaches them from the narrower perspective of energy and the so-called woodfuel "crisis" of Africa and other parts of the Third World. It does so for two basic reasons.

First, woodfuel and related energy problems are important and pressing topics in their own right. Since most Africans are poor and can afford or have access to little other than firewood, charcoal, or crop and animal residues to meet their basic energy needs, woodfuels dominate the energy economies of virtually all African countries. In sub-Saharan Africa they account for 60–95 per cent of total national energy use, with the highest proportions in the poorest countries and in the household sector, even though consumption is small by world standards and amounts roughly to only one cubic metre of wood per person annually, or a mere quarter of a ton of oil equivalent. It will take many years of rising incomes and infrastructure development before such countries can afford alternatives to this massive woodfuel dependence.

The more obvious symptoms of this dependence are well known. In many places woodfuel resources are dwindling because of deforestation which is caused to varying degrees by the need for farming land, and by over-grazing, commercial logging, uncontrolled fires and tree cutting for fuel. As wood resources diminish and recede, for millions of people the costs of obtaining woodfuels, whether in cash or time for gathering them, are imposing severe and increasing strains on already marginal household survival and production strategies. These impacts are greatest for the poor and for women, who normally bear the responsibility for fuel provision and use. They are not yet felt everywhere, but they are spreading to more places and more people.

Great efforts will be needed to reduce these impacts, prevent them from spreading, and provide sustainable and adequate energy supplies at affordable costs for fast-growing populations.

The second reason for adopting a woodfuel perspective is that ideas about the nature of the woodfuel "crisis" and what to do about it are beginning to change quite fundamentally, with far-reaching implications not only for energy specialists and decision makers but for all the disciplines and institutions which are concerned in some way with the land, environment and sustainable livelihoods – from agriculture, forestry and rural development to urban planning and systems of law governing land-use rights.

Over the past fifteen years, large policy, planning and donor aid structures have been created and hundreds of millions of dollars committed to addressing the woodfuel crisis directly as a problem of *energy* supply and demand. The issues appeared quite simple and the remedies self-evident. Where trees and woodfuels were "scarce" or getting "scarcer" many direct solutions seemed to offer a good chance of quickly bringing forests and woodlands, woodfuel supplies and woodfuel demand into balance. This would both improve welfare by reducing the costs of obtaining fuels and "save the trees".

Governments and foresters set about trying to protect public forests and woodlands from encroachment by woodfuel cutters. Foresters tried to increase woodfuel supplies with peri-urban plantations and village energy woodlots. Energy ministries tried to curb rising consumption by promoting more energy-efficient cooking stoves, or reduce pressures on the forests with more efficient charcoal kilns. Attempts to promote the use of oil and electricity instead of woodfuels became key elements of African and other Third World energy strategies.

While there have been a few successes with these energy-focused efforts, most have failed to turn the tide of wood depletion or prevent growing pockets of fuel scarcity. But as one expensive disappointment has led to another and simple certainties have begun to evaporate, important lessons are now being learned.

Better information is showing that many of the most basic assumptions on which these efforts were based are false or highly misleading: for instance, that the use of woodfuels is normally the principal cause of deforestation, or that the expanding circles of deforestation around cities will inevitably force up woodfuel prices and hence provide a powerful economic rationale for all kinds of afforestation and conservation measures.

At the same time, it is now increasingly recognized that by focusing so closely on woodfuels and the symptoms of their scarcity, these direct approaches looked only at the tip of the proverbial iceberg and ignored the much broader and deeper strains in the environmental, social, economic and political fabric of which woodfuel scarcity is only one manifestation. They obscured the fact that woodfuels are only one of many basic needs and that their provision – for example, by "tree growing" – is only one aspect of household coping strategies and land-use systems on the farm or in the village. Indeed, these top–down and over-specialized approaches often failed to notice that in many places rural (and urban) people were already responding to woodfuel and other land-use stresses in ways that are imaginative, innovative and with far lower cost than most project interventions.

The more comprehensive and objective view of woodfuels which is now emerging recognizes that there are no single, simple answers and that the problems surrounding them are inseparably linked to the complex, diverse, extremely dynamic and multi-sectorial issues under-lying Africa's broader crisis of population, food, poverty, land and natural resource management. Equally, successful remedies for wood-fuel problems must be firmly rooted in these broader contexts. In particular, if planning, projects or other types of intervention are to create lasting successes they must recognize at least three basic factors:

1. the need for local assessments and actions and the unhelpful nature of large-scale averages. The "landscapes" and "peoplescapes" of Africa, especially, are extremely diverse. Problems and opportunities to solve them are therefore *specific* to place and to social groups in each place. The aim should be to reach underlying causes rather than heal the symptoms

2. the need for *indirect* approaches to woodfuel issues and greater *participation* by local people at every stage to help them to prioritize and solve their own problems. This follows from the first point, and also from the fact that success normally depends on starting and strengthening processes rather than delivering technical packages: on "how" rather than "what" things are done
3. the need for decentralized and multi-disciplinary approaches, including the use of competent and trusted "grassroots" agencies, to facilitate the two first points. However, this does not exclude the need for economic, legal and political initiatives at the macro-level to improve the broad contexts for local, positive change.

One might well say: so what's new? Isn't all this now broadly accepted? Well, accepted maybe – but not yet acted upon. Although these perspectives are in tune with the broad paradigm shift which is now sweeping through governments, aid agencies and other parts of the development community, one has to bear in mind the enormous inertia and vested interests which can resist such basic changes to conventional structures of authority, responsibility and knowledge.

If we restrict ourselves to woodfuel issues, it is clear that energy policy makers, planners, analysts and project staff need to redefine their roles, learn new concepts and styles of working, and even surrender their bureaucratic empires and specialist corners to others. New institutional linkages, joint policies and data gathering, and other kinds of alliances between government agencies, extension services and the like – as well as new forms of interventions – are all needed if woodfuel issues are to be addressed in a more holistic and relevant way.

Much the same is true of foresters, who often justify their efforts as direct remedies to the woodfuel crisis while failing to recognize the broad contexts which underlie them. Agricultural and livestock schemes, in turn, have often ignored the needs of foresters (and local people) and have often greatly worsened local woodfuel problems by ignoring them as a planning issue. Narrow specialism, false diagnoses of problems and top–down attitudes are found in all of the many disciplines and institutions which work, directly or indirectly, towards the better management of land and natural resources. Getting off the beaten track and heading for new territory with unfamiliar allies will not be easy either for them or for woodfuel specialists.

The main purpose of this book is to support these changes by showing why they are necessary from a woodfuel perspective and hence what they imply for resolving woodfuel problems. However, it should be obvious from the remarks already made that this aim will take us far deeper into many of the fundamental issues to do with the broader environmental and production crises of Africa and what might best be done about them.

Of course, the book cannot cover every aspect of such a vast subject. Nor can it always point with certainty to effective solutions. Everyone concerned with these large issues is on a steep learning curve: as yet there are few cut and dried "answers". What the book does attempt is to review as objectively as possible the main issues, positive options and constraints (as well as areas of profound ignorance) and successful achievements to do with the woodfuel problem in its much broader setting of sustainable land use and natural resource management. It uses actual case studies wherever practicable to back general discussions with "on the ground" realities.

However, one thing that the book does not try to do is cover all issues of "energy for development". Its focus is on woodfuels as basic needs: "energy for survival". We feel that dealing with this must take a higher priority than looking beyond survival needs to conventional energy developments or the plethora of renewable energy devices such as biogas plants and windmills that have been so widely proposed as aids to development, especially in the rural areas of the Third World.

This first chapter lays the groundwork of the book by making a sometimes harsh critique of conventional views of the woodfuel "crisis". It does so in order to ask the basic question: what is the woodfuel problem? Without a clear understanding of what the problem is, or what the underlying causes are, there is little chance of developing appropriate and effective policies and other interventions to meet it.

WOODFUEL GAPS AND THE DEATH OF THE FORESTS

The woodfuel "crisis" of developing countries was "discovered" in the mid 1970s when much of the world was gripped by the energy crisis of modern fuels which followed the first oil price shocks of 1973–4. The scale of deforestation across the Third World was already recognized. As energy analysts and anthropologists began to pile up the evidence across the developing world about the huge scale of woodfuel use and the difficulties that millions seemed to be facing in getting enough wood as tree stocks declined it seemed natural to regard both types of crisis as essentially similar.

The woodfuel problem seemed to be a classic case of rising energy demand outstripping supply. Although the resources in this case were renewable – unlike oil, gas and coal – they were apparently being overused at unsustainable rates. So a numbers game known as woodfuel "gap theory" was conceived which quickly came to dominate almost every attempt to measure the scale of both the woodfuel crisis and the remedies which would be needed to alleviate it.

For instance, all of the sixty-odd UNDP/World Bank energy sector assessments for African and other developing countries which considered woodfuels in the first half of the 1980s adopted gap theory methods. They were used for the extremely influential UN Food and Agriculture Organization (FAO) study in the early 1980s which estimated that in 1980 just over 1,000 million people were living in areas of woodfuel "deficit" because they were cutting tree stocks to meet their energy needs faster than the trees could regrow, and that this number would almost double to 2,000 million by the year 2000.[1] The same idea underpins many more recent and widely quoted reviews of Africa's forestry and woodfuel problems[2-5] and was being used even in 1988 by major donor agencies to justify large-scale forestry projects in Africa.

The basic premise of gap theory, as normally practised, is that woodfuel consumption is the principal cause of deforestation and therefore of mounting woodfuel scarcities. To measure the scale of this impact, one estimates the consumption of woodfuels (and sometimes of timber, construction poles and other tree products) in a given region and compares it with the standing stocks and annual growth of tree resources. The latter may be scaled down to allow for controlled forest reserves, game reserves, and trees in remote places where access is difficult.

Typically one finds that consumption greatly exceeds the annual growth of trees. For instance, studies of the Sahelian countries found that woodfuel use exceeded the growth rate of tree stocks by 70 per cent in Sudan, 75 per cent in northern Nigeria, 150 per cent in Ethiopia and 200 per cent in Niger, with a small surplus of 35 per cent in Senegal.[4]

The next step is to project these present-day gaps. Since consumption has to be met from somewhere, one assumes that the difference – the "gap"– is made up by cutting into tree stocks. Woodfuel consumption is projected, usually in direct proportion to population growth, and calculations are made of the resulting tree stock each year. As consumption rises and trees are felled, the annual growth falls, the gap grows bigger, and the tree stock is still further depleted. Inevitably, the

Table I.1: Woodfuel Gap Forecasts for Sudan

	1980	1985	1990	1995	2000	2004*
	(million cubic metres of tree stock)					
1. Forest stock	1,994	1,810	1,539	1,145	607	57
2. Forest growth	44	40	34	25	14	1
3. Woodfuel consumption	76	88	102	121	141	159
4. Woodfuel gap (3 – 2)	32	48	68	96	127	158

Note:
* Extrapolated from published data to year 2000.
Source: Anderson and Fishwick (1984).[2]

stock of trees declines at an accelerating rate towards a final woodfuel and forestry catastrophe when the last tree is cut for fuel.

Table I.1 shows the results of just one of many applications of this method, in this case for Sudan.[2] Forest stocks will fall to zero by 2005. A similar exercise for Tanzania, published in 1984, showed that the last tree would disappear under the cooking pot by 1990.[6] There are still many trees in Tanzania.

The final step is to ask what must be done to close the gaps and bring consumption and tree resources into balance. With few exceptions, the answer is afforestation (or demand management by the dissemination of more efficient cooking stoves, etc.) on a staggering scale. For instance, the World Bank study which did much to legitimize woodfuel gap theory estimated that tree planting in sub-Saharan Africa would have to increase fifteen-fold in order to close the projected gaps in the year 2000.[2] The vast scale of these remedies, and the calamitous consequences if they are not applied, naturally tend to combine to provide strong justifications for large, centrally directed, plantation forestry projects focused on woodfuel provision.

In criticizing these methods we do not intend to imply that there are not serious and growing woodfuel shortages in many places; that woodfuel consumption does not often exceed renewable supplies; that afforestation is not, for many reasons, an admirable objective; nor even that supply-demand analysis, of which traditional gap theory is one model, is not a legitimate tool for resource assessments at the national or regional level.

Our criticisms concern the serious practical and theoretical flaws in gap theory as it has so often been (and still is) applied. By ignoring these flaws, gap methods have done much to exaggerate the scale of the woodfuel problem and foster inappropriate, large-scale, energy-focused remedies at the expense of other actions which could have done much more to improve welfare, reduce deforestation, and generally support sustainable development.

One serious flaw is that the large-scale aggregate perspectives of gap theory help to obscure the fact that woodfuel problems are location-specific and require precisely tuned and targeted interventions, usually on an individually small scale.

The second flaw is that this numbers game is played with weak numbers. While this fault is widely acknowledged, the game continues and its conclusions continue to be taken with great seriousness. Three points deserve emphasizing:

- estimates of woodfuel consumption and of tree resources are rough and in many cases little more than guesses, yet it is the relatively small initial difference between them which drives the gap forecast. This difference is extremely uncertain, thus putting the whole projection in serious doubt
- estimates of tree stocks are particularly uncertain and are usually gross underestimates of the resources which are actually used for fuels. Most such statistics are held by forest departments, which typically know little or nothing about the volumes of trees outside the "forest" – for example, on farm lands, fallow lands and village commons – or about scrub, bushes and other forms of non-tree woody biomass resources. These additional woodfuel resources may be very large
- most gap predictions assume that once a hectare of forest has been cut it becomes "dead land", without any natural regeneration of trees or shrubs. Adding in this factor of tree regrowth can soften dramatically the dire predictions of gap forecasts.

A third and more fundamental flaw concerns the forecasting methodology. As noted on p.6, consumption is usually assumed to rise in line with population, even while supplies dwindle to vanishing point (see Table I.1, p.7, for example). Everyone acknowledges that this is unrealistic and that as scarcity worsens and wood prices or the labour costs of gathering fuels increase, many coping strategies would come into play. Tree planting would increase, consumers would use fuels more economically, or they would switch to more abundant fuels such as crop residues, and so on.

Nevertheless, these inevitable responses are usually disregarded because *no one knows how large they will be*. In economic jargon, there is virtually no information for woodfuels (or tree growing) on the price elasticities of supply or demand. Any downward adjustments that the gap forecaster made to the rising consumption curve would therefore be entirely arbitrary. Consequently, it is better to leave it alone and present the gap predictions as "trends continued" scenarios which are designed only to show what large adjustments must be made to demand or supply to bring them into balance.

The crucial flaw to this apparently reasonable attitude is that it greatly exaggerates the need for *planned* interventions. It implies that all the supply-demand adjustments must be *implemented* when in fact many of them will be made (and are already being made) "naturally" by ordinary people without any external assistance. This fault could be corrected by better information, thus putting gap theory on a respectable footing; but until this is done the method must be regarded as a dangerously misleading assessment and planning tool.

Another most fundamental flaw of gap theory, as it is so often used, is its basic assumption that deforestation is driven mostly (or in many models, entirely) by woodfuel consumption. We turn to this question next.

WHERE DO WOODFUELS COME FROM?

How would all gap theory predictions change if woodfuels were only a minor cause of deforestation? In particular, what if most woodfuel supplies arise as a by-product of agricultural land clearances?

If this is the case, woodfuel gap forecasts are facing the wrong way. Rather then being driven by woodfuel demand, they should be driven by data and trends about land use, from which woodfuel supplies arise merely as a by-product.

Many other aspects of the woodfuel "crisis" are also turned on their head. To arrest deforestation one needs to halt the depredations caused by agriculture rather than by woodfuel consumption. Measures to reduce woodfuel use become much more a matter of improving welfare by cutting consumer costs than attempts to "save the trees". Indeed, if all woodfuel use stopped tomorrow, deforestation rates would hardly be altered. Most importantly, the main strategy for halting deforestation would be to intensify cropping and grazing systems so that less new land is needed as populations grow. This puts the onus of maintaining forest cover and woodfuel supplies firmly on the agricultural system – or agriculture plus forestry – rather than "energy".

There would also be major implications for the economics of forest use. It is often said that urban woodfuel consumers pay too little because their supplies are "mined" from the forest with only trivial payments (if any) for the use of these public resources. In turn, this leads to the argument that substantial royalties should be paid by woodfuel cutters for the use of state-owned trees, to reduce pressures on them and to pay for their intrinsic value as well as their less obvious environmental and other social benefits.

Now there may be several good reasons for (as well as severe difficulties in) introducing such measures. At this point we simply note that if land clearance is the main cause of deforestation it is the farmer, not the woodfuel trader, who should pay the resource costs. Since much of this clearing is by subsistence farmers, or smallholders living on the edge of survival, collecting substantial resource royalties from them would be, to say the least, difficult. We return to these and related questions in Chapter 7.

Equally important, if land clearing is the major cause of deforestation the strategic woodfuel questions change. The key question for the sustainability of woodfuel supplies is now: how long can land clearing for agriculture – with its surplus woodfuel bounty – continue before new land "runs out"? And if the rate of clearing slows as agriculture is intensified, what will this do to woodfuel supplies? How might the supply-demand system adapt? Some interesting answers to these questions can be found in South Asia, where the land clearing frontier was reached some time ago and where most woodfuels now come from managed trees on farm and village land rather than the forest.[8]

Is this also the future for Africa? In Chapters 1 and 2 we look at the many positive gains to be made by rural tree growing and management, in Chapter 3 at the constraints which can prevent them, and in Chapter 4 at the institutional innovations and linkages which can help to overcome these constraints.

Unfortunately, there is little robust information to answer the critical question of where woodfuels do in fact come from. The best that can be said is that there are several major sources and that the relative importance of each appears to vary greatly from place to place. Each place – particularly each city – must discover the facts for itself in order to plan realistically and effectively. However, of the five main sources of woodfuels described below, there is strong evidence that the last two are in general, on the large scale, by far the most important.

1. *Tree cutting directly for fuel*, especially to make charcoal. This certainly occurs around some cities and in more distant patches of forest close to main roads or rail. Cutting may be intensive, amounting almost to clear felling, but is usually more selective so that only the larger trees of suitable species are felled to leave a fairly complete cover of smaller trees. This may be left to regenerate or, since the charcoalers have made the work easier, cleared for farmland. One can see large areas of degraded woodland like this around Blantyre, Malawi, for example. The larger trees have gone, leaving a dense patchwork of smaller trees interspersed with patches of maize and vegetable crops. The sustainability of this source depends on whether trees are replanted and on cutting rates compared with the rates of natural regeneration in the affected areas.

Although these operations can be a well-organized, round-the-year business, it is important to appreciate that in many places they reflect failures in the agricultural system. Much commercial firewood and charcoal destined for the cities is produced from trees felled by rural people to supplement their incomes, especially in the slack agricultural season, or in years when returns from farming are poor due to drought or low farm prices. To give just one example at this point, tree felling for sale as woodfuel by Communal Land farmers in Zimbabwe greatly increased during the years of drought and crop failure in 1978–81 but then returned to normal levels.[9]

2. *Dedicated woodfuel plantations*. Although these are quite common in Asia and there is interest in them as sources of urban woodfuels in Africa, supplies in Africa are at present small except for a few cities such as Addis Ababa, Ethiopia, where woodfuel prices are exceptionally high. As we shall see in Chapter 6 and 7, their economic viability depends mostly on whether single-purpose "industrial" or mixed small-holder methods are used and on the price structures of urban woodfuel transport and markets.

3. *By-product wood* from various tree growing activities: for example, "lops and tops" from multi-purpose farm trees, or commercial forestry for timber, or specialized farm tree crops such as gum arabic in Sudan and tannin from small woodlots of wattle in Kenya and several southern African countries. As discussed later, multi-purpose trees and other forms of woody biomass are crucially important sources of woodfuels for rural people; whether or not they can provide urban supplies depends on the complexities of urban market structures and prices.

4. *Dead branches and twigs* which are picked off the ground or cut from the tree. Many surveys confirm that these non-destructive sources account for the great bulk of rural firewood which is taken from state-owned forests and woodlands as well as a good deal of the wood from managed tree and other woody resources on farm lands and village commons.

5. *Surpluses arising from agricultural land clearances*, including rotational fallow or "slash and burn" farming systems. As discussed below, these sources usually greatly exceed local woodfuel needs, even though many trees may be left standing as part of the farming system, while others are burned to provide soil nutrients or simply to clear the land. The extent to which they provide woodfuels for the cities again depends on prices and market structures. Much of the forest devastation which surrounds urban centres must have been due to agricultural land clearance rather than direct cutting for fuel, since it is normally much more profitable to use land within reach of urban markets to grow food for the city than to leave it under trees and sell the wood. The pressures to clear such land of trees, sell any salvage wood to the city, and then farm it, are almost irresistible.

How important is land clearance as a cause of deforestation? When he was chief forestry adviser to the World Bank, John Spears estimated that from 1950 to 1983 it was responsible for about 70 per cent of the permanent forest destruction in Africa. A rough calculation can take this estimate further by showing why land clearance must be a major source of woodfuel. In sub-Saharan Africa, average cropland is close to 0.4 hectares per person.[10] Lands which are typically cleared for farming, such as savannah woodlands and reasonably productive bush, have standing tree stocks of about 20–30 cubic metres per hectare.[11] So with population growth, and without any increase in per hectare farm yields, each extra person's food needs have to be met by clearing some 8–12 cubic metres of wood, or enough to meet a typical annual per capita woodfuel consumption of 1–1.5 cubic metres for five to twelve years. Even if some trees are left standing, or are burned, or one assumes standing stocks of only 7–10 cubic metres per hectare typical of degraded savannah or dry bush, there will be substantial surpluses to provide for woodfuels.

More direct evidence of the crucial role of land clearing comes from survey data. In Botswana a rural energy study carried out for the UK

Overseas Development Administration[12] made extensive surveys in the eastern region and concluded that the rate of tree felling was many times greater than consumption of the principal wood products: firewood and building poles. Most trees were felled to clear farmland, when they were commonly burned to waste. The second most common reason for felling was for building poles. Fuelwood in rural areas came mostly from fallen branches and the smaller pieces salvaged from land clearances.

In Zimbabwe, a large and detailed survey of wood use, tree planting and forest cover based on aerial photographs in the Communal Lands gave unequivocal confirmation that land clearance has been the main cause of tree loss over the past two decades.[13] In every one of the six classes of land distinguished by their average forest cover, field areas have increased, mostly at the expense of degrading medium-forested land (25–50 per cent tree cover) to woodlands with less than 25 per cent cover.

In Zambia, a recent study concluded that deforestation is primarily caused by agricultural practices.[11] Although trees are cut for firewood and charcoal around the major "line of rail" urban centres, the main deforestation problems arise from a combination of population pressure and the failure to intensify farming systems. In Luapula, Eastern, Southern and Northern provinces over 20 per cent of forest land has been cleared for farming. This process is probably accelerating, since the more successful farmers are increasing production and incomes by acquiring more land, not by intensifying their production methods.

In Sudan, a progressive improvement in data about the wood energy system has led to an almost complete reversal in perceptions about the causes of deforestation.[14] The large National Energy Assessment (NEA) completed in 1983 concluded that the national loss of trees amounted to 75 million (solid) cubic metres a year and that 95 per cent of this was due to woodfuel consumption (62 per cent charcoal and 33 per cent firewood). Over the next few years serious faults in this analysis were revealed by the Sudan Renewable Energy Project which was co-funded by the Sudanese government and the US Agency for International Development. Careful surveys showed that charcoal production was nearly twice as efficient and consumption much less than previously supposed. Consequently a large project to improve the efficiency of charcoal making was hastily abandoned, since the traditional producers were already getting higher yields than expected from much more expensive and sophisticated technologies. The NEA had also made a large error when converting woodfuel consumption measured in stacked cubic metres to solid cubic metres "on the tree".

When these faults were corrected, estimates of the national tree loss were reduced more than three-fold to around 22 million cubic metres a year, the proportion due to woodfuels fell to 72 per cent, and the contribution from farm clearances and shifting cultivation rose tenfold from 2.3 to 23 per cent. However, if firewood is excluded from these estimates, since little of it comes from tree felling, the changes are even more dramatic. The proportion of the tree loss due to charcoal is now only a third, while mechanized land clearing and shifting cultivation account for 55 per cent. Although these new data are not the last word in accuracy, they undoubtedly come closer to reality than the earlier estimates. Table I.2 summarizes these changes and gives a very different picture from the woodfuel consumption and deforestation rates shown for the gap prediction for Sudan in Table I.1 (p.7).

Striking evidence for urban woodfuels comes from Kenya, where in 1986 a survey was made of the sources of charcoal for four urban centres. It was found that for Nairobi and Nakuru nearly all the charcoal came from clearance of forest or rangeland for agriculture, and the remainder from sustainable wattle and other plantations. For

Table I.2: Changing Views on the Causes of Deforestation in Sudan

	original estimates (NEA 1983)		revised estimates (Gamser 1988)	
	(million cubic metres of tree loss per year)			
Firewood	24.94	(33%)	13.13	(59%)
Charcoal	46.53	(62%)	2.93	(13%)
Land clearance, etc.	1.75	(2%)	5.04	(23%)
Over-grazing, fire, poles, etc.	2.13	(3%)	1.12	(5%)
Total tree loss	75.35		22.22	
Excluding firewood:				
Charcoal	46.53	(92%)	2.93	(32%)
Land clearance, etc.	1.75	(4%)	5.04	(55%)
Over-grazing, fire, poles, etc.	2.13	(4%)	1.12	(13%)
Total tree loss	50.41		9.08	

Notes:
NEA = National Energy Administration.
Land clearance = mechanized agriculture plus shifting cultivation.
Source: Gamser (1988).[14]

Kisumu 80 per cent came from the latter sources. Mombasa's supplies came from a mixture of land clearing and low-intensity felling on forest or rangelands.[7]

As we have said, such findings cannot be generalized: diversity is the rule and the only valid information comes from specific data on the conditions in each country and, more importantly, each district and city. Yet it is clear that for these places at least, deforestation and attendant woodfuel problems depend mostly upon the growing demands for farm and grazing land rather than for woodfuels. Woodfuel supplies are mostly a residue of these pressures or come from more or less sustainable systems of multipurpose tree management elsewhere.

Where this is the case, the obligation for slowing deforestation and maintaining woodfuel supplies at affordable costs falls on a wide range of sectors. Apart from such basic issues as reducing population growth, the main sectorial strategies are:

- for agriculture, to intensify production per unit of land. Much broader issues than deforestation and woodfuels are obviously involved here, including the improvement of food security and rural incomes. Equally, there is a wide range of possible strategies, from conventional technical approaches such as mechanization, irrigation and fertilizers, to the many "softer" approaches which are now being developed and are presented in Chapters 1 to 4
- for forestry, to support many of the softer approaches to agricultural intensification and, more generally, to enhance rural livelihoods, by various "agroforestry" approaches. These are also discussed in Chapters 1 to 4. Trees in farming systems provide vital environmental services, especially the protection and improvement of soil and water resources, as well as supplying essential products such as animal fodder, construction materials, food, medicines and fuel
- for forestry, to counter deforestation by growing trees or managing natural forests wherever this is economically justified. The economic sums need to include a wide range of environmental concerns, from the fact that trees fix carbon and help to limit global warming due to the release of carbon dioxide, to watershed and soil protection, the maintenance of genetic diversity and support of wildlife, as well as the benefits and costs of supplying valued wood and other tree products. Providing jobs can also be

an important economic rationale for afforestation schemes
- for energy, to support these strategies indirectly since they help to maintain woodfuel supplies. Otherwise, most energy options centre on demand management in order to increase the welfare of consumers by reducing their energy costs.

Where woodfuel needs are a direct cause of deforestation, all three sectors obviously need to add the provision of woodfuels at affordable costs to these agenda.

GIVING SCARCITY A HUMAN FACE

One of the most basic assumptions of conventional woodfuel thinking is that physical scarcity of wood is the key issue to address. As typified by gap theory, analysts and planners have measured the scale of woodfuel problems in terms of volumes of wood resources and consumption and distances from resources to consumers. In rural areas, the distance and time taken to collect woodfuels are commonly used as the main yardstick of scarcity and the need for remedies. For urban areas, it is commonly assumed that woodfuel prices will rise as forest stocks are depleted and the transport distances from the city to its main woodfuel resources lengthen. Since increasing physical scarcity or distance can impose considerable costs to consumers, the basic aims of woodfuel interventions are to reduce these costs by reducing physical scarcity.

There is, of course, a good deal of truth in these attitudes. However, interventions are most unlikely to succeed if they do not recognize that physical scarcity means nothing unless it is related to the human dimension. As Peter Dewees[15] has cogently pointed out, one has to ask whether these costs are the outcome of physical scarcity itself or of much more fundamental issues such as labour shortages, land endowments, social constraints on access to wood resources, or cultural practices. These "human issues" are both complex and dynamic and frequently undergo rapid and adaptive change which the outsider may easily miss.

Consider wood gathering, for instance. There is now compelling evidence from time budget studies for rural women that the time spent collecting firewood can vary greatly from one week or season to the next depending on agricultural and other labour demands; it is often minor even in "wood-scarce" areas compared with time for collecting water, food preparation, cooking and other survival tasks; and it is both adjusted and perceived as a more or less severe problem, in relation to the totality of labour needs and time available.[16,17] Similarly, it is the total time required for cooking and fuel collection which is one of the critical factors deciding whether women burn "scarce" but otherwise preferred and easily managed species of firewood or turn to more abundant and easily collected fuels such as crop residues or animal wastes which may take more time and trouble to use in the cooking fire.

The basic issue is therefore one of labour availability, not fuel availability. If spare labour is abundant it may not matter if woodfuel-collecting trips are long or getting longer. If labour is very scarce, even the collection of abundant woodfuel supplies may be perceived as a serious problem. What matters is local perceptions of these questions and the coping strategies that people have evolved or are evolving to deal with them, not the outsider's simple physical measurements.

Labour scarcity on the farm may also be a deciding factor in rural tree growing versus crop production (see Chapter 3). It is even an important factor in setting urban woodfuel prices. It is commonly assumed that these will rise mainly because "scarcity rents" will be imposed as trees are depleted, and transport costs will rise as haulage distances increase. But as we shall see in Chapter 6, urban woodfuel prices depend to a considerable extent on the abundance or otherwise of agricultural labour through its effects on the costs of harvesting trees or making charcoal. Many factors other than the physical scarcity of trees are also critical, including the availability of trucks and the demands made on them for all purposes; shortages (and prices) of alternative "modern" fuels; crop prices and their effects on rates of land clearance; urban wage rates and employment prospects; and the degree of competition in the woodfuel market. Many of these factors have changed and could be changed substantially in order to increase the incentives for urban woodfuel production on a sustainable basis without planting a single tree to reduce wood "scarcity".

This brief discussion has hardly scratched the surface of what physical "scarcity" actually implies in human terms. All that we have tried to do is suggest that many issues are involved and that the diagnoses of woodfuel problems and the design of remedies for them have to reach these underlying dimensions. Many ideas on how to do this, with case histories and a more careful discussion of the issues involved, are presented in the following chapters.

To conclude these introductory comments we present in Figure I.1 a

Figure I.1: The Contexts of Rural Woodfuels

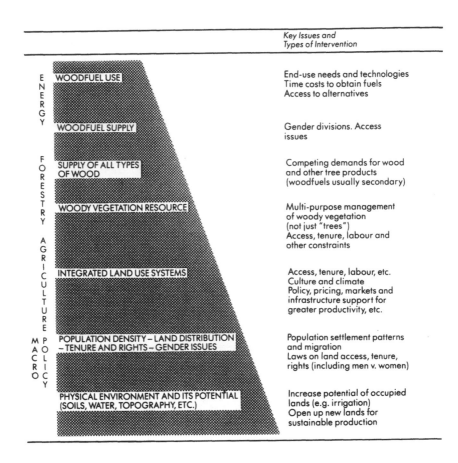

Key Issues and
Types of Intervention

WOODFUEL USE
End-use needs and technologies
Time costs to obtain fuels
Access to alternatives

WOODFUEL SUPPLY
Gender divisions. Access
issues

SUPPLY OF ALL TYPES OF WOOD
Competing demands for wood
and other tree products
(woodfuels usually secondary)

WOODY VEGETATION RESOURCE
Multi-purpose management
of woody vegetation
(not just "trees")
Access, tenure, labour and
other constraints

INTEGRATED LAND USE SYSTEMS
Access, tenure, labour, etc.
Culture and climate
Policy, pricing, markets and
infrastructure support for
greater productivity, etc.

POPULATION DENSITY – LAND DISTRIBUTION – TENURE AND RIGHTS – GENDER ISSUES
Population settlement patterns
and migration
Laws on land access, tenure,
rights (including men v. women)

PHYSICAL ENVIRONMENT AND ITS POTENTIAL (SOILS, WATER, TOPOGRAPHY, ETC.)
Increase potential of occupied
lands (e.g. irrigation)
Open up new lands for
sustainable production

ENERGY
FORESTRY
AGRICULTURE
MACRO POLICY

sketch which underlines our basic point that woodfuel issues must be seen and acted upon in their broadest contexts. It makes two main points.

First, conventional woodfuel approaches have usually looked only at the top of the diagram, using simple energy-centred measures to decide whether there were significant problems demanding interventions. It is now increasingly recognized that a more holistic and participatory approach to problem identification, diagnosis and intervention must be used. Whether or not people feel that woodfuels are a significant "problem" must depend on their views of the costs involved compared with the many other concerns and costs in their lives. If they do not recognize these costs as significant, then a woodfuel intervention that itself has significant costs compared with benefits, or does not simultaneously reach other and more urgent concerns, will in all probability be rejected or allowed to collapse into failure.

Second, on this more holistic view, woodfuel issues are merely the tip of a seamless pyramid which reaches down to progressively more fundamental aspects of survival, production and land management. Various disciplines – energy, forestry, agriculture, land use planning, and the like – have each staked out layers of this pyramid as their territory. Again, it is increasingly recognized that each discipline must do much more to look at and link with the layers both above and below its own special concerns. At the same time, one needs to look sideways, across the pyramid, at the various issues and conflicts which are found at each level.

Only a highly integrated and multi-disciplinary perspective can hope to succeed in this formidable task. That is why, at the local level, a highly participatory approach is essential. The people who live closest to each of these differing, local pyramids are the people who know most about them and who are most likely to recognize – albeit with external assistance of various kinds – the key leverage points for beneficial change and the brakes which could prevent it.

However, this participatory "bottom–up" approach is not sufficient because it has little to do with the many constraints and possible levers for positive change which are found at the base of the pyramid: the macro-level issues which are the concern of top–down planning and government policy. The strengthening and co-ordination of policies and actions across these higher levels to match the realities and opportunities at the micro-level is an extraordinarily difficult task but is also one that must be faced if sustainable development is to be achieved.

Part I

Rural Areas

Chapter 1

Trees for Rural People

Over the last two decades there has been a major shift in attitudes to forestry and rural tree growing by the donor community and governments in Africa, as views have changed about what problems these activities were supposed to address. Although the shift is by no means complete, today there is a broad consensus that smallholder tree production for multiple purposes is perhaps the single most important option for local people, donors and governments in easing rural woodfuel problems and – even more importantly – in helping generally to sustain rural livelihoods and reduce environmental risks.

At its leading edge, this change in attitudes has been quite remarkable. Until the early 1970s, almost all development forestry followed a Northern model of large-scale industrial-commercial plantations and natural forest management to earn revenues and foreign exchange. The needs of rural people were largely ignored, as were trees outside the forest. Foresters and agriculturalists hardly talked to each other, and neither paid much attention to the vital roles of trees within farming systems.

By the mid 1970s, people-centred approaches to rural development were beginning to gain ground and foresters were slowly drawn into them. This period saw at least the first signs of some interaction between foresters, agriculturalists and social scientists within the development community. In 1978, for example, the FAO initiated its influential Forestry for Local Community Development Programme.

By the end of the 1970s and early 1980s, farm and community forestry was in vogue. Training courses were developed, new institutions were established, and many projects based on various kinds of social forestry – including agroforestry and wood energy plantations – were started. In the mid 1980s some major reviews of these experiences were published[1,2] and forestry was increasingly integrated with agriculture and other rural development efforts as donors became more aware of the need to address the multiple priorities, constraints and opportunities of local people. At least, these ideas were increasingly reflected in the rhetoric of development agency reports and conference papers, if not in actual activities on the ground.

As an indicator of this progression, the World Bank, for example, has in the past decade devoted over 60 per cent of its forestry budget to supporting smallholder tree growing (in twenty-seven forestry and forty agricultural and rural development projects) whereas in the previous decade 95 per cent of its forestry spending went on industrial or large-scale plantations. Other major agencies have matched this shift of resources towards agroforestry, farm and community forestry. From a minimal level of funding twenty years ago, at least eleven major international groups now help to finance such projects in more than fifty countries. According to the International Council for Research in Agroforestry (ICRAF), agroforestry projects are now under way in at least 100 Third World countries.

This part of the book is intended to support this progression and move it still further towards the idea that, to be successful, rural development must be *by* rural people rather than *for* them. All too many forestry projects are still designed for rural people in the offices of northern capitals or the five-star hotels of the South. Phrases like "participation" and "meeting multiple needs" have entered the rhetoric of forestry development but are widely misunderstood and rarely acted upon. "Top–down" attitudes based on the notion that "we" know best still prevail. In the context of woodfuels, narrowly conceived wood energy forestry projects costing many millions of dollars are still designed and implemented based on "fuelwood crisis" thinking for regions where they do not meet local needs and the money and effort could be far better spent in more broadly based activities.

This chapter sets the argument going with a deliberately upbeat view of the capabilities of rural people to manage their own development, either unaided or with some outside help. It focuses on tree growing and the management of woody vegetation through a number of examples taken from actual experience. Some of these concern spontaneous initiatives by farmers and livestock herders, based on their priorities and the wealth of indigenous technical and environmental knowledge which many rural people possess. Most of the examples, though, look at a spectrum of initiatives by outside agencies which were designed to assist such farmer-led activities, ranging from the most participatory or "bottom–up" approaches to some which began in an authoritarian "top–down" manner but made little headway until they changed their approach. The chapter concludes with a look at the economics of rural tree growing which emphasizes the need for some radical rethinking by conventional foresters when they come to deal with rural tree growing and management.

Chapter 2 then takes a more formal look at the most important tree-related technologies which have been or could be employed in the great diversity of African land management environments. These are the technical options or tools for improving rural tree growing and management for multiple benefits or for overcoming the more serious local constraints. While we do not advocate a technical package or "tool kit" approach – since technical thinking alone misses most of what is important about rural livelihood systems – it makes sense that, whatever the ecological or socio-economic conditions, the technical package is a good one. Indeed, it should be the best that collaborative research and development (between outside researchers and development practitioners, and local people using "popular science") can come up with.

These two chapters might give the impression that local knowledge, a little sensitive outside assistance, and the impressive tool kit of "agroforestry" are all that is needed to win the battle to grow more trees as a vital ingredient of rural development. Chapter 3 shows why the battle will be a long one. It examines the many constraints and conflicts of interest which are commonly found in rural communities and which prevent successful tree growing activities. These include the multiple "rural squeeze" of scarce labour and cash, the denial of access in many forms to land and other resources, problems of tenure and rights to land and trees, and divisions of interest between men and women.

The most critical development question of all is *how* to go about addressing these constraints rather than *what* technical packages one might apply to them. This is the theme of Chapter 4. Here we do not enter into a discussion of detailed planning issues. Instead we outline the salient characteristics of the general *approaches* which seem most promising. In doing so we draw on a number of practical examples of land management interventions and institutional innovations which have addressed the fundamental constraints governing access to wood products for rural people and their ability to improve that access, for example by growing trees. These examples tell of some notable successes. But they also convey a sense of the awful constraints and the great diversity and complexity of African rural livelihoods and environments which, together, make the topics covered in this book so urgent, difficult and challenging.

POPULAR KNOWLEDGE AND EMPOWERMENT

Various terms have been used to describe the base of local knowledge which rural communities apply in managing their environment, such as community environmental knowledge[3] or indigenous technical knowledge.[4-6] This local knowledge base is often rich. It is always specific to local landscapes and livelihood strategies.

The intricacy of indigenous technical knowlege (ITK) systems reflects the diversity of local agro-ecological conditions and the importance this has for livelihood strategies geared to managing risk. For example, several researchers have documented sophisticated local soil classifications based on soil colour and texture and their relevance for local cropping patterns.[7] Livelihood strategies themselves are correspondingly complex and diverse, since this is crucial to their sustainability in risky environments.

Community environmental knowlege may have accumulated over many generations, but since it is gained from experience it encompasses only those aspects of the local environment which are important to people's livelihoods. Rural people are not *all-knowing* about *all* aspects of their landscape. Pastoral communities, for example, tend to have much more comprehensive knowledge of the many uses to which local trees and shrubs may be put, than of how actively to grow them.

Rural livelihood systems are also dynamic. They are open to changes brought about by farmers and herders who are actively engaged in "social livelihood experimentation".[8] Indeed, agriculture has been likened to a *performance* played out by a specific farmer – a skilled actor – on a specific piece of land in a specific year,[9] applying his or her knowledge in different ways to solve land management problems as they arise.

Given all this, it is no surprise that many of the most successful interventions for natural resource management are successful precisely because they *support* this creative process of social livelihood experimentation, or at least take ITK as the starting point in rural development. Several researchers have made this case eloquently and stressed the need for "farmer participatory research" as a basis for appropriate interventions.[4-6, 10, 11]

In brief, the rationale for supporting indigenous knowledge, social livelihood experimentation, and the empowering of local communities and individuals to direct change, is threefold.

First, local communites take a holistic view of their livelihood systems. They do not reduce them to the narrowly defined disciplines of agriculture, forestry, livestock husbandry, health care and nutrition, and to "problems" within them. The local, holistic view is more likely to address the causes of these problems – which Robert Chambers has referred to as "integrated rural poverty"[5] – rather than merely their symptoms.

Second, given the complexity and diversity of local land management systems, outside researchers and development agents alone could not possibly identify, and still less implement, appropriate actions in all contexts. Farmer participatory research makes more sparing and cost-effective use of scarce scientists and other outsider skills.[12]

Third, farmer participation is more than simply a means towards more effective local development; it is also an ideological goal in itself, in the development of local capabilities, self-confidence and power to control one's resources for betterment.[13]

Limits to Local Knowledge

However, one must also recognize that there are limitations to local knowledge systems and people's ability to apply them successfully on their own. Rural people do not always have useful knowledge to contribute in all circumstances. Their scope is obviously limited to the local pool of techniques, materials and genetic resources. Furthermore, local knowledge systems rarely, if ever, have either the necessary forward perspective or resources to anticipate the opportunities and constraints arising from changing environments.[14] Local knowledge is also rarely distributed uniformly throughout a community. The capacity of individuals to generate, implement and transfer ITK will vary. One reason is the economic stratification which is found in virtually all rural societies. On the whole, individuals with greater assets or income are likely to innovate the most, although poorer people may be forced to innovate because of their poverty.[11,15] In short there may be many constraints which farmers cannot overcome without external support.

Indigenous technical knowledge also requires a favourable social context for its successful implementation. One outstanding condition is community stability.[11] "Stability" does not imply "resistance to change", for just as rural livelihood systems are dynamic, so are the communities in which they are embedded. They are complex and flexible and their resilience should not be underestimated. New community groups can and have been formed, such as self-help groups and women's groups.

But there are several ways in which community stability can be disrupted to prevent effective change. One is the pressure of commodification – the penetration of a cash economy – which often leads to sacrifice of the common good for short-term individual gain. Second, population pressure on resources and incipient land degradation in many places are exceeding the capacity of local institutions to adapt and manage environmental change.

Third, and perhaps most importantly, local institutional capacities are often overwhelmed and eroded by the pace of change. As a result the poor and powerless in particular have often lost access to productive resources, as outsiders and powerful local interest groups consolidate their control over them.[16] This often leads to the resource "mining" of forests and rangelands, traditionally sustainably managed common property resources, to provide capital for investment outside the system or in concentrated individual holdings. Common property resources – and those dependent on them – are generally the first to suffer, owing to a reliance for their management on more or less fragile social institutions. In the agropastoral livelihood systems of the dry Kenyan savannas for example, poor households often depend on nearby off-farm or shared lands for 50–90 per cent of their fodder and fuelwood needs,[16] while community controls and stability have been weakened:

> ...to implement ITK in respect of, for instance, communal grazing without irreversible resource depletion, requires a respect for the common good – characteristic only of strong social organisations. Where these have broken down, vestiges of ITK relevant to resource management may survive, but will not necessarily be relevant to new organisational forms that will have to be developed with external assistance.[11]

Finally, on a different but related point, one needs to recognize that there are strong biases against farmer participatory approaches both among national governments and the international donor community. One reason is that these demand a reversal in the way "normal professionals" think about rural development, with their tendency towards increasing specialization and standardization of approaches in disciplines such as forestry, agronomy and livestock husbandry. This point is now becoming more widely appreciated but the challenges it presents to "development professionals", in both the North and the South, should not be underestimated.[17] A second major reason is that many small, local, participatory activities – none of them with any clear-cut guarantees of "success" or neat budgets and cost-benefit tags attached to them – are simply more difficult for donor agencies and governments to fund, manage, staff and claim credit for than large multimillion-dollar projects.

FARMER-LED INITIATIVES

In 1987 a team from Energy/Development International undertook a major assessment of successful natural resources management initiatives in the West African Sahel.[18] The team looked at over seventy different initiatives in four countries – Mali, Niger, Senegal and Gambia – many of them by local people without the benefit of any outside interventions. This is how they describe what they encountered:

> The team looked at the West African Sahel as a cup half full, not half empty. Everywhere we saw people substituting their own imagination, technical skills, physical energy and knowledge for what were once the naturally occurring gifts of rain, dependable crop yields, grazing pastures, forests, rivers and groundwater, game, fish, and medicinal plants. Persistent drought and economic crisis have changed West Africa permanently. One can look at the massive deterioration of the environment and predict doom, or one can look at the bold and creative acts of individual small farmers, fishers, and herders and see what a difficult, but not impossible, future might bring.[18]

Under the right conditions, rural communities can take all kinds of spontaneous collective action to arrest environmental degradation and to improve natural resource management. Success stories do exist, as the following examples will show, and although they may be the exception and not the rule, it is important that the common features of such initiatives are explored. These commonalities can be brought to bear on outsiders' attempts to support participatory interventions elsewhere, in similar agro-ecological contexts.

The right conditions for indigenous initiatives to work appear to include: (1) heightened local perceptions of environmental degradation and its negative welfare impacts; and at the same time (2) a degree of political organization within the community to facilitate effective management. The latter may depend on relatively stable social relations but is not necessarily confined to the most socially homogeneous communities. For example, those in the Mwenezi district of Zimbabwe (see Chapter 5, Rural Case 2) are markedly heterogeneous but are still able to invent effective resource management practices.

The Shinyanga region of Tanzania (Rural Case 1) and the Mwenezi district of Zimbabwe (Rural Case 2) provide examples of varied spontaneous local actions in the face of land degradation and a declining wood resource base. Both examples centre on the importance of livestock in these agropastoral livelihood systems. Maintaining large herds of cattle may appear to outsiders to be incompatible with the aim of increasing woody plant cover, but total destocking would lead to an intolerable level of risk – or complete loss of livelihood – especially during periods of drought.

In Shinyanga, farmers and pastoralists have responded to increasing pressures on local wood resources by developing ways of propagating trees and shrubs for live fences from cuttings, transplanting wild plants and direct seeding rather than conventional nursery seedlings which are hard to obtain. Offtake from the live fences, which are designed to control grazing and allow fenced-off areas of vegetation to regenerate, is also used for fuel. Several communities have developed sophisticated rules and sanctions to regulate access to woody vegetation and traditional grazing reserves.

In Mwenezi, local people have taken a lead in establishing an improved system of land management which clearly demarcates settlement, arable and pastoral areas, with rotational grazing paddocks marked out with multiple-purpose live fences. As a result, cattle have acquired a much greater commercial value, local woody vegetation has increased for a variety of uses (including fuelwood), and new orchards have been planted to increase food supplies and cash incomes.

Both examples demonstrate that improved pasture management, and the restriction of grazing to limited areas which this makes possible, permit both an improvement in livestock quality and an increase in tree and shrub cover more generally. Local recognition of these potentials has enabled village councils to enforce effective community sanctions against those who break the rules, since the rewards are plain to see.

Broadly similar innovations have been made by the Pokot pastoralists in the Baringo district of Kenya, who have developed a carefully controlled system of regulated access to traditional grazing reserves. A form of transhumance is practised, with livestock being moved gradually from the lowland wet season grazing areas to the hill areas in the dry season where the perennial grass is to be found. Within these areas grazing reserves are set aside for use, at the discretion of the elders, in hard times.[19,20]

In the much drier Turkana district of Kenya, a similar system of range management is practised, but on a larger scale, with movement over greater distances (see Rural Case 3). Grazing is carried out under a co-operative grazing community known as an *adakar*, which represents a semi-permanent cluster of homesteads which come together in the wet season.[19] The Turkana also have a strong tradition of sorghum cropping, and as an adaptation to drought have developed the fastest-maturing known variety of sorghum, which requires only sixty days in the ground from planting to harvesting.

In Senegal, an exceptional group of villagers came together in 1974 after the first wave of recent Sahelian droughts to form an association of self-help village development groups known as the Koumpentoum Entente (Rural Case 4). With funds generated from an innovation in collective field management they have gone on to develop a wide range of natural resource management initiatives. In the village of Diam Diam, for instance, several different kinds of reafforestation are being undertaken by the Entente sub-committee on some 250 hectares of village land set aside for the purpose. Youth groups and women's groups are involved in the planting activities. A particularly interesting innovation is the establishment of individually owned livestock herds from a nucleus of co-owned goats and cattle. A scheme for fertilizing fields through a carefully planned system of rotating livestock corrals is also underway in Diam Diam. The various Entente projects are separately managed and each is responsible for raising the necessary funding from outside bodies, but control remains in the hands of the villagers themselves.

But perhaps the most remarkable examples of innovation to meet changing conditions are found among individual farmers, where community cohesion and stability are not really an issue. Such examples could be drawn from all over Africa, but we have chosen two from the same broad region of south-western Kenya to illustrate the extraordinary diversity of practices and skills which farmers can develop when pressed to do so.

In the high potential agricultural areas of Kisii and Kakamega districts in Kenya, population densities have reached very high levels indeed. Land is scarce and farms have been repeatedly subdivided. Over large areas there is scarcely any natural woody vegetation left. Under these conditions, however, farmers are responding to stress in highly imaginative ways, making very intensive use of limited space and biomass resources.

Ainea Mundanya, for example, has a farm which is a mere 1½ hectares in size in Kegoye sub-location of Kakamega district. He keeps three dairy cows, stall-fed with a fodder mix prepared on the farm. The fodder combines dried and chopped Napier grass with dried leaves from the *Calliandra* trees on the farm, or with pods from his *Sesbania* trees. These leguminous trees are intercropped with the Napier grass in a highly diverse home garden, and were grown from seedlings obtained from the Kenya Woodfuel Development Programme (KWDP; see Rural Case 9). Mundanya has also grown a live fence of *Leucaena* trees to mark his farm boundary. He meets a substantial part of his fuelwood needs from tree prunings on his farm. He has a seedling nursery which supplies his own needs and enough over to sell to neighbouring farmers. He also sells seeds back to KWDP. Manure from the cattle stall is carried around the tiny farm in 2-gallon drums to sustain soil fertility under the leguminous trees and fodder grass.[21]

In Kisii district, Joseph Mogaka has a plot of about 0.8 hectares which he farms so intensively that, as he says, "we don't allow the soil to cool down". On three tiny fields he is developing woodlots for fuel and fodder, with black wattle seedlings (*Acacia mearnsii*) in one, *Sesbania* species in another, and plantings of mixed species in a third. A fourth field is being planted with Napier grass as fodder for his three cows. Another field of Napier grass is interplanted with leguminous fodder species (*Mimosa* and *Calliandra*). In other areas maize is interplanted with leguminous trees in the corner and *Mimosa* on the boundaries, and there is a mixed vegetable plot. There is a field for permanent grazing, and another growing pyrethrum for sale which is interplanted with maize, leguminous trees, bananas and beans. Mogaka wants to buy two cows not just to produce milk for home use and sale but because of a complex and carefully judged understanding of the need on such a small and intensively used farm to maintain soil fertility to sustain yields. Fallows are short, the cow dung now produced is not sufficient for all the land, and crop residues which would make excellent fertilizer have to be fed to the cows. The new cows could eat leaf fodder from trees which are now maturing and would produce the extra dung he needs.[22]

Both Mundanya and Mogaka are constantly experimenting and making changes in their farming strategies, applying the technical knowledge they have learned from experience together with new ideas gleaned from wherever they find them.

SUPPORTING LOCAL INITIATIVES

Where local people cannot themselves overcome the constraints they face in land management, success depends on support from outside. In this section we consider a number of initiatives which have achieved notable success in improving land management practices. Their common theme is outside support seeking to build on existing local knowledge and institutions on the grounds that this is vital for success and sustainability. The only way in which the majority of rural problems can be solved is by willing action on the part of local people and by helping to empower them to use their own knowledge and skills to take charge of their own development.

The examples range from farmer-led initiatives with minimal outside support, to outsider-led initiatives where efforts have been made to work with and address people's felt needs and priorities, and imposed projects. In other words, they cover the full spectrum from "bottom–up" and participatory interventions to "top–down" and authoritarian projects. All the examples involve rural tree planting or tree and shrub management, since these "agroforestry" or community forestry approaches are more likely than conventional forestry to benefit the poor and be broadly accepted. In all but the most "top–down" approaches, local knowledge either formed the basis of project design or was used to alter a project's direction and methods once it was under way.

Table 1.1 summarizes the external inputs provided and the potentials for improved resource management which were realized in cases where these can be defined. The real meat and flavour of each example can only be extracted by reading the rural cases themselves (see Chapter 5). In the text that follows we try to do no more than give a taste of each intervention and the problems it tried to meet.

We start with two cases which we have already discussed as examples of spontaneous innovations without any external inputs.

Table 1.1: Interventions in Wood Resource Management:
External Inputs Provided and Local Potentials Realized

Example	External agency	External inputs provided	Local potentials realized
Konso Development Programme, Ethiopia (Rural Case 5)	NORAD/ NORAGRIC	• Food-for-work initially • Transfer of nurseries to local people	• Capacity of peasant associations to manage tree planting • Local knowledge of indigenous tree species and agroforestry
Research in the Mazvihwa, Zimbabwe (Rural Case 6)	ENDA-Zimbabwe	• Development of a process to solve local problems systematically: – Community seminars to discuss wood problems and potentials – Short training courses in nursery techniques and community resource management	• Extensive indigenous ecological knowledge of local woody plant resources • Capacity of local communities to draw up woody plant management plans
Chitemene farming in Zambia (Rural Case 7)	Soil Productivity Research Programme of Zambian Min. of Ag. with NORAD and ICRAF	• Application of D&D methodology to identify additional agroforestry options to increase wood supply • On-station research into improved fallow systems	• Farmer participatory research into improved fallow systems • Better use of trees for mulch
Alley farming and dairying in Kenya (Rural Case 8)	Ministries of Energy and Regional Development; Agriculture and Livestock Development	• Two independent programmes providing: – Research station agroforestry package (alley cropping) – Cattle and dairy improvement	• Farmers' ability to modify agroforestry package to suit own needs • Holistic local view of livelihood system enabling co-ordination of the two externally initiated programmes
Kenya Woodfuel Development Programme (Rural Case 9)	KWDP	• Extension programme to change attitudes to women's tree planting • Mass awareness programme using popular theatre, etc.	• Improvement of tree and shrub management on farms by enabling women to participate more effectively
Yatenga Agroforestry Project, Burkina Faso (Rural Case 10)	OXFAM	• Simple water tube levels, and training of farmers in their use • Maize flour loan scheme to permit resource-poor farmers hire help to build water-catching bunds	• Farmers' indigenous technical knowledge of stone bunds to catch water • Local abilities to adapt technology • Collective organization of labour to build stone bunds in individual fields
Community forestry projects, Sudan (Rural Case 11)	ERC with government and Sudan Renewable Energy Development Programme	• Funds • Technical advice on forestry, if requested	• Local capacity to draw up and implement effective land use management plans

In the Shinyanga region of Tanzania, the HASHI soil conservation programme (Rural Case 1) is responding reactively to the numerous instances of spontaneous initiatives in resource management mentioned above. The programme will take an innovative approach in building on traditional management practices such as creating grazing reserves and protected areas to allow the regrowth of woody vegetation, not to mention the sophisticated techniques of tree and shrub propagation which local people have developed.

In the driest region of East Africa, a rural development programme funded by Norway has been working with the Turkana pastoralists to help them develop their remarkable system of individual ownership and management of riverine forests known as *ekwar* (Rural Case 3).[23] The *ekwar* plots of *Acacia tortilis* and *doum* palm forest provide fruits, construction materials and medicines, and are an extremely important source of dry season fodder and browse for livestock. Due to recent droughts and increased pressures from charcoal makers created by a new tarmac road, the *ekwar* system is declining and the forests are degrading. The forestry component of the project is attempting to revive the system by maintaining the best of local knowledge and management skills but also combining them with modern forestry and management techniques. However, its method is very much a "joint learning" approach based on seminars with the local village leaders and government foresters and extension services. The project is also achieving in this harsh, dry area some remarkable results in terms of tree growing and natural recovery of vegetation by small-scale water harvesting and voluntary restrictions on the areas where cattle can graze.

In another dry area, this time in South Ethiopia, the Konso Development Programme (KDP) appears to be making real progress with a broadly similar joint learning approach in which local institutions are empowered to manage woody plant resources (Rural Case 5). Supported by Norway, the original objectives of this integrated community development programme with a tree planting component were to reduce soil erosion and provide fuelwood and building poles. However, in this semi-arid area where livelihood security depends on both rain-fed agriculture and herding, local people were concerned much more with fodder adequacy and soil fertility than with other uses for trees, particularly those which would yield benefits only in the longer term.

The project took a commendably flexible aproach and revised its plans accordingly. Over time, outside involvement shifted from the initial use of food-for-work incentives to get people to start planting trees, and radically revised the methods of supplying planting materials. The use of distant, state-run seedling nurseries was abandoned in favour of local facilities run by peasant associations. The latter were also given full control over tree planting. Above all, the considerable local knowledge of indigenous tree species and agroforestry practices was put first in developing a site-specific, participatory approach to wood resource management.[24] The KDP clearly has faith in the capacity of local people to innovate and experiment; this is essential if they are to feel committed to its activities. Most importantly, it makes the initiative far more sustainable and promotes the capacity to take *other* initiatives, when the project itself ends.

The fourth example comes from the Mazvihwa district of Zimbabwe, where the non-governmental organization ENDA-Zimbabwe has facilitated a community-based woodland management and extension project (Rural Case 6). Local people recognized that new land management problems and stresses had emerged which swamped their own capabilities. The project took as its starting point the extensive body of indigenous ecological knowledge, and organized community seminars for the open discussion of problems and potentials in woody plant management. With administrative back-up and co-ordination from ENDA, the local people then prepared their own wood resource management plans. It is hoped that this new initiative will support local efforts in tree planting and the control of grazing in communal areas.

The next case concerns the *chitemene* system of shifting cultivation which is widespread in northern Zambia (Rural Case 7). In its traditional and extensive form, this long-established farming method has become unsustainable in places where a growing population and the increasing commoditization of the rural economy have combined to exert pressure on local forest resources. The Soil Productivity Research Programme of the Zambian Ministry of Agriculture and Water Development has been exploring ways of building on farmer-led experimentation into improved bush fallow systems, in order to develop land management systems which can sustain higher population densities.

With support from the Norwegian Agency for International Development (NORAD) and ICRAF, a Ministry of Agriculture team identified opportunities for collaborative research and development with local farmers, particularly by building on the farmer-developed *fundikila* system of soil mounding and composting which has substantially increased the population carrying capacity in some areas. Using ICRAF's Agroforestry Diagnosis and Design methodology, ways of better integrating planted trees into the system have been identified so as to enrich woodland fallow and provide nutritious leaf mulch for the mounding process.

Another example is one where innovations by farmers, based on their priorities, radically altered two outsider-led technical agricultural projects (Rural Case 8). An alley cropping package, in which trees are interplanted with field crops, was developed for farmers in the coastal region of Kenya by the Mtwapa Agroforestry/Energy Centre to increase soil fertility and food crop yields. But on leaving the research station the technical package was modified by farmers to emphasize livestock-fodder production rather than food crop yields; it became an alley *farming* system. A Dairy Development Programme (DDP) in the same area, under the Ministry of Agriculture and Livestock Development, was designed and implemented independently of this agroforestry system. It fell to local farmers to cure this lack of "official" institutional co-ordination and identify the missing links between the two programmes. The problem of fodder adequacy had not been fully appreciated by the DDP and it was up to local farmers to see that they could supply fodder from their newly adapted alley farming system.

In the same country, the KWDP (Rural Case 9) has for long worked on the view that in the areas of high agricultural potential and population where it operates, access to woodfuel is a problem of socio-cultural rather than physical constraints. The KWDP's community extension programme uses innovative techniques such as popular theatre and community meetings, together with mass rallies, to enable people to see for themselves the reasons behind their woodfuel problem. In large part the explanation lies in gender divisions over access to managed woody vegetation, combined with variable opportunities for marketing wood products such as poles (by men), which take priority over the use of wood to meet household fuel needs (by women). KWDP wants to create an awareness of the problem – which currently affects women – as a *family* problem. As a way of improving access to wood for fuel they are promoting tree species which are relatively fast growing but which do not yet have a cash value attached to them, so that their produce stays on the farm rather than going for sale in local towns.

OXFAM's Agroforestry Programme in the Yatenga region of Burkina Faso (Rural Case 10) is now a widely documented "success story" in supporting local land management initiatives. However, it is worth repeating here some of the lessons learnt from the programme. The project initially encouraged farmers to plant trees using a water harvesting technique adapted from a system which is used in the Negev desert of Israel. In time, however, it became clear that farmers were more interested in the effects of rain water harvesting on food crop yields. In response, the project changed tack and developed a simple water harvesting innovation – stone contour bunds – which was an extension of an existing local practice for intercepting rain water runoff. By means of a simple farmer training course, and farmer-to-farmer diffusion of the technology, the stone line technique spread rapidly and led to impressive increases in food crop yields. Institutional support was also provided by the OXFAM project. A system of maize flour loans was set up to enable resource-poor farmers to enlist the help of their neighbours in the labour-intensive task of constructing bunds on their fields. This institutional support builds on an existing local organization: the collective work group.

Local perceptions of the severity of land degradation in Yatenga were undoubtedly a major factor in the enthusiastic response of farmers to the stone line innovation. Nevertheless, much of the credit for the programme's success must go to the project staff, whose flexible approach and attention to local needs ensured the constructive participation of local people.

Several community forestry projects in the Kordofan region of northern Sudan (Rural Case 11) have been facilitated by an innovative approach which was adopted by the Sudan Renewable Energy Development Project and the Sudanese Energy Research Council (ERC). This was for villagers to design, implement and manage their own forestry projects, with support from the ERC in the form of funding and technical advice, if needed. This working partnership between villagers and national forestry authorities proved to be an extremely effective way of allowing local people to solve their own problems. The villagers of Um Inderaba have established a tree nursery; planted and protected a tree windbreak; fenced off a small area to allow for the regrowth of woody vegetation; and planted trees for shade, fuel and fodder.

Significantly, the most successful local initiatives were taken by the remotest communities, despite harsher physical conditions. In villages where resident outsiders provided forestry support, local initiatives actually appeared to be stifled. Most success was achieved where maximum responsibility for development activities lay in the hands of the villagers themselves.

Finally, as a contrast to these highly participatory interventions we present two examples where "top-down" approaches have, at best, delayed successful development.

The first is the Majjia Valley Windbreak Project in Niger, which has taken a decade or so to become a now widely documented "success story" (Rural Case 12). With support from the non-governmental organization CARE and the Niger Forest Service, it began in 1974 when villagers approached a local forester for help in planting tree lines to protect their fields from severe and worsening problems of wind (and water) erosion and loss of vegetation cover. Since then almost 500 kilometres of windbreak trees have been planted, with dramatic benefits for biomass production, fuelwood supplies and crop yields, the latter especially in the worst drought years (see Table 1.3, p.46).

However, although the project always intended to treat local farmers and herders as partners and principal beneficiaries, some "top–down" approaches in its early years caused considerable hardship to some groups of local people and have contributed to the fact that it has taken a decade or so for most villagers to accept the project and co-operate fully with it.

In the early years the windbreaks were located by project staff and planted using food-for-work incentives. The trees were protected by paid guards, and large fines were imposed if animals were caught among the growing trees. Herders, and women who lost grazing rights for sheep and goats, were the principal losers. Farmers were also resentful because land was appropriated for the windbreaks without any compensation. Over time, though, the obvious benefits from the windbreaks combined with a more participatory approach by the project itself have resolved many of these difficulties, leading to a high degree of local involvement. For instance, local people are now deciding for themselves on the location of new windbreaks (which are also spreading spontaneously) and are applying their own agreed sanctions over guarding and excluding people from windbreak and grazing areas.

The second example is a much more extreme case, in which state power was used to enforce the removal of cattle from pastoralists who depended on them for survival. Although the intentions were entirely for the best – to conserve the soil and improve livelihoods in a badly degraded region – it was carried out despite the strong opposition of local people. The example is the HADO soil conservation project in the Kondoa Eroded Area of Tanzania (Rural Case 13).

It is an interesting case because despite the high-handed approach, the destocking did achieve significant benefits and local pastoralists did learn to adapt to it. Furthermore, in both this aspect of the project and in a linked series of village afforestation efforts which are also described in the Rural Case, the project has gradually adopted a more participatory approach and now appears to be achieving greater success.

The question which both the Majjia Valley and HADO examples raise is whether the technical packages they pursued would have taken off faster and achieved greater and more rapid benefits for the environment and local livelihoods if there had been a more participatory and co-ordinated planning process from the outset.

Ingredients of Success

Several factors are common to this handful of initiatives (and of others which are presented in later chapters) which have been chosen from many possible examples. One is that success, and particularly continuing success, is more likely if one incorporates and builds on indigenous technical knowledge. Placing control in the hands of local communities has also been the key to success in several examples, where external agencies have provided only limited inputs but have tried to empower the communities themselves to manage the process of change. But perhaps the most important lesson of all, in terms of the institutional process behind the interventions, is the need for external agencies to be flexible and responsive to local needs, even where this requires a complete reorientation of approach.

The main ingredients of success, then, are for outsiders to go in quietly, with eyes and ears open, and to have the humility and patience to learn, change course if need be, and take a long-term approach.

THE ECONOMICS OF RURAL TREES

The most obvious point which arises from this brief review of tree-management practices and interventions is that success cannot easily be measured, least of all simply by the numbers of economics. Success – and the ensuing benefits – depends on such intangibles as a growing enthusiasm, or self-confidence, or community spirit which empowers people to develop technical packages or systems of management for themselves to meet their own needs and priorities. Starting or fostering such human processes is infinitely more important than any attempts to quantify the tangible outputs of such beneficial change.

Nevertheless, measurements of benefits (and costs) are important. They are, after all, what farmers notice and talk to each other about. They are also important to donor agencies, government departments and other institutions involved in rural development who need to justify their actions, make choices about future actions, and confirm where they have made the right (or wrong) choices on what to do.

In this section, therefore, we examine some of the arguments for rural tree growing as seen from "above" – from the perspective of outsider agencies – focusing on tree growing and management on the farm. This is followed by a final section which looks at how farm forestry demands some new ways of economic thinking which may be strange to conventional foresters who have been brought up in the traditions of plantation and forest management.

Low Costs

One clear advantage of farm forestry over conventional plantations is its low cost. Comparisons are difficult to make because of differences in soils and climate and also in the items which are included in project costs, and hence in costs per tree. Equally important, farm forestry is normally part of a broader package of project inputs than the conventional plantation and typically provides multiple benefits besides "timber". Comparing like with like in these circumstances is not easy.

Nevertheless, some broad comparisons can be made which demonstrate the generally enormous cost advantages of farm forestry. According to one review the average cost per seedling for conventional forestry projects in the Sahel during the 1970s was around US$11, although this includes all project overheads.[25] However, the average survival rate of seedlings was a mere 2 per cent, giving a cost per surviving tree of some US$580. During the 1980s costs for similar projects in the Sahel have fallen slightly and survival rates have improved to around 10 per cent, but the average surviving tree still costs about US$100 to establish, when all project overheads are included.

By contrast, some recent farm forestry projects in Africa have produced seedlings for well under US$1. For example, seedlings in the

CARE-Uganda Village Forestry Project cost only US$0.11 each,[26] and in the CARE-Kenya Agroforestry Project at Siaya about US$0.07.[27] School nurseries in Zimbabwe operating alongside the Co-Ordinated Agricultural and Rural Development Project can produce seedlings for only US$0.01 each.[28] Although these figures are for seedlings rather than trees (as in the Sahelian examples in the last paragraph), seedling survival rates are frequently much higher in farm forestry than in conventional forestry projects. In a number of African farm forestry projects the costs of a surviving tree are less than US$1. For example, even in the harsh conditions of the Sahel, the SODEVA experiment with *Acacia albida* in Senegal produced seedlings for US$0.88 each and, because subsidies were paid only for surviving trees, the survival rate was virtually 100 per cent.[29] This gives a cost per surviving tree of under US$1, better value for money by an order of magnitude than even the best of conventional forestry projects at present. However, not all operational farm forestry projects in the Sahel have been so successful. CARE's *Acacia albida* extension projects in Chad, for example, have reported seedling survival rates as low as 5 per cent, and in no case higher than 70 per cent.[30]

To reduce costs further, there is a trend towards propagation by seed rather than seedling, at least in high potential areas, and towards nurseries run by farmers (or schools) rather than forest departments and other "outsider" organizations. New low-cost methods of seedling care are also being tried. The heavy-gauge black polythene tubes and imported insecticides which are commonly used in conventional nurseries are rarely essential for raising village forestry species: lower-gauge polythene and locally available wood-ash termicide would be perfectly adequate. Other low-cost techniques which need to be tried alongside the decentralization of nurseries include seedlings raised in baskets or earth-balls.

A recent evaluation of alternative forestry strategies in Tanzania bears out these points.[31] The study went much further than the usual simple economic comparisons of tree planting and of woodlot plantations by advocating and costing broader, mixed approaches to tree management to provide the same products – in this case fuelwood and construction poles.

The study concluded that the net gains from the present village forestry strategy are small (or even negative) owing to the high cost and low returns of raising and distributing seedlings from centralized nurseries, using imported, heavy-gauge polythene tubes and insecticide.

Instead, much greater production at lower costs – and therefore far higher net returns – could be had from a mixed strategy of plantations based on decentralized nurseries and the regeneration of natural woodlands using direct sowing. The far greater productive potential and economic benefits of this approach compared with the conventional strategy are compared in Table 1.2. The analysis also underlined the importance to costs and output of involving the villagers in protecting their trees. This would also serve to clarify the issue of who has ultimate responsibility for the rights over the trees and tree products; namely, the villagers.

Table 1.2: Comparison of Costs and Benefits from Alternative Village Forestry Models for Tanzania

	Model I *current practice*	*Model II* *proposed practice*
1. Cost per seedling raised (1986 Tanzanian shillings)	1.54	1.15
2. Seedlings planted per hectare	1,110	1,600
3. Fuelwood produced in 24 years (m³)	800	3,200
4. Poles produced in 24 years (number)	6,000	10,500
5. Financial rate of return to villagers, assuming all activities undertaken by them, but no soil "protection" costs, and products sold at market prices	11.2%	21.4%
6. Economic rate of return	– 4.0%	11.8%
7. Economic rate of return (without protection costs)	2.7%	23.6%
8. Soil conservation activity	nil	as required

Assumptions:
Model I: 5-hectare plantation (or existing demonstration woodlot) with a central nursery based on existing practices. Total area: 5 hectares.
Model II: 5-hectare plantation, plus 1 hectare individual planting and 2-hectare sowing of blanks in village's own natural forest. Decentralized nursery close to village based on proposed practices. Total area: 8 hectares.
No staff or overhead costs in either model.

Source: Adapted from Swedforest Consulting AB (1986).[31]

In a similar vein, the World Bank has recently presented an economic case for woody biomass management rather than subsidized seedling distribution for Kenya.[32] In Kenya the Forest Department presently concentrates on seedling production rather than providing technical advice on improving management practices, and bears heavy financial losses by doing so. The department sells 60,000 seedlings a year at a price of 7.5 shillings per 100, although it cost 50 shillings per 100 to produce them. The resulting annual subsidy of 26 million shillings represents 7.5 per cent of the department's total budget. The World Bank has recommended that subsidies be abolished and that the Forest Department seedling production should be confined to the ecological zones and species for which growing conditions are more difficult – including trees in arid and semi-arid lands, and species which are difficult to germinate on farms, or have low survival rates, or for which seed stocks are small.

Higher Farm Yields

What of the production potential of rural tree growing? As noted above, this is hard to measure because, in most cases, the outputs and benefits of farm forestry are so diverse. Its aim is to use the productive and protective features of woody plants to increase, sustain and diversify total output from the land since it is this, rather than the production of a single "good", which matters in the livelihood strategies of resource-poor farmers.

In Chapter 3 we look at the conditions where these positive objectives cannot be met because of ecological, social and economic constraints. If "agroforestry" was always the magic road to riches which some of its enthusiasts appear to claim, there would be less need for the immense efforts that are now being devoted to fostering it. Indeed, there are many people who are unable to take it up – or who can actually lose by it.

Yet despite these very real problems, many studies and trials of various agroforestry combinations have shown that they can provide large benefits not only in terms of extra wood but also as fodder for livestock and increased yields of crops. The fact that agroforestry can increase total output from the land, and in particular increase food crop production, is one of the most powerful arguments in its favour, as food tends to be people's first priority. Indeed, sometimes quite small changes in management practices which involve the use of trees in farming systems can lead to quite remarkable increases in "bio-economic potential".[33]

For example, *Acacia albida* trees which are found throughout much of the Sahel, shed their leaves just before the rains and the main season for crop growth. The leaves are ploughed into the soil to increase fertility, while crops growing under the trees are no longer affected by their shade and do better than those grown in the open. In Senegal, Charreau and Vidal estimated the fertilization of the soil by *Acacia albida* leaves to be equivalent to 50 tonnes of manure per hectare each year in stands of fifty trees per hectare.[34] Annual millet production was doubled from 500–800 kilograms to 1,000–1,500 kilograms of grain per hectare as a result. In addition, the average tree produced an annual 400–600 kilograms of pods, giving a theoretical stocking rate of 1.5–2.0 sheep or 0.15–0.20 head of cattle per hectare from this form of fodder alone.[35] Interestingly, however, *Acacia albida* on the Kenyan coast does not shed its leaves in the rainy season, possibly because of a higher water table,[36] and so is less valuable for this type of farming system.

Research in Rwanda has demonstrated that stall-fed livestock can gain from around 15 to 20–30 grams per day when a ration of grasses is supplemented with leaves from woody plants. The greater weight gain is largely a result of the higher protein content of the leaves, which is typically 20–26 per cent for dry leaves compared with less than 5 per cent for grasses.[37] In the Kakamega district of Kenya a few farmers have begun experimenting with such fodder combinations for "zero-grazing" (stall-fed) cattle under conditions of severe land shortage. A successful fodder combination is to mix cut and dried Napier grass with the dried pods of *Leucaena* species or the dried leaves of *Calliandra*.

Table 1.3 summarizes some of these results, together with the results of other trials and studies which bear witness to the increases in crop yields made possible at least partly by interplanting with trees. At the same time there may be very considerable offtake of wood for fuel use. Table 1.3 includes such data where they are available.

The demonstration effect of such trials and projects may be sufficient in some cases to stimulate the spontaneous replication of the relevant agroforestry practices in neighbouring areas. It is not difficult to secure farmer support for an agroforestry strategy which gives demonstrable and fairly immediate increases in crop yields or fodder and hence in food security. This has been observed, for example, in the Majjia Valley Windbreak Project in Niger (Rural Case 12) which has consistently shown crop yield increases of around 20 per cent (and large potentials

Table 1.3: Crop Yield Increases from Agroforestry Trials, Studies and Projects

project/trial/study	tree species	food crop	initial crop yield (kg/ha/yr)	increased crop yield (kg/ha/yr)	percentage increase (%)	potential firewood yield
1. IITA alley-cropping trials, Ibadan, Nigeria On-farm trials in same area	Leucaena	maize	1,040	1,900	83 39	6–7 (tons/ha/yr)
2. CARE Agroforestry Extension Project, Kenya (alley-cropping trials)	Leucaena	maize	1,000	1,750*	75	5–10 (tons/ha/yr)
3. Intercropping trials with food crops and Gmelina sp. in Malawi	Gmelina arborea	beans groundnuts maize	1,024 1,605 9,804	1,292 1,616 9,373	26† 1† –4†	8–10 (m³/ha/yr)
4. CARE Majjia Valley Windbreak Project, Niger	neem (Azadirachta indica)	millet/ sorghum	—	—		110m³ per km (standing stock)
• Windbreak protection trials			396.5	488.0	23	
• Fertilizer application trials			397.0	487.5	23	
• Control plots versus unprotected plots			—	—	15–60†	
• Pollarded windbreak lines versus uncut controls			—	—	97–537‡	
5. Acacia albida intercropping on farms in Groundnut Basin, Senegal (infertile sandy soils)	Acacia albida	groundnuts/ millet	500 (±200)	900 (±200)	80	
6. OXFAM Yatenga Agroforestry Project, Burkina Faso	(water- harvesting)	millet/ sorghum				
year: annual rainfall: 1981 692 mm			510	857	68	
1982 421 mm			442	495	12	
1983 413 mm			295	418	42	
1984 383 mm			153	292	91	

Notes:
* Assumes 25% land taken out of crop production by windbreak lines.
† Not statistically significant.
‡ Statistically significant.

Sources: Bognetteau-Verlinden (1980),[38] Edje (1982),[39] Felker (1978),[40] Harrison (1987),[41] Jensen (1987),[42] Kang et al. (1985), [43] Nair (1984), [44] Nkaonja (1985), [35] Reij (1987), [45] Rorison and Dennison (1986), [46] Vonk (1987).[47]

for fuelwood and timber offtake) as a result of the protective effects of the windbreaks. In another case shown in Table 1.3 – the use of *Acacia albida* trees in the Groundnut Basin of Senegal – the yields of crops growing near the trees on these infertile sandy soils were around 80 per cent greater than those away from the trees.[40] However, we should point out that the impressive yield increases of 12 to 91 per cent shown in Table 1.3 for the Yatenga Agroforestry Project (Rural Case 10) are the result of water-harvesting techniques rather than agroforestry.

The recent *Kenya Forestry Subsector Review* drew attention to the quite staggering benefits that can be obtained from farm forestry.[32] For example, it estimated that sales of poles and building timber from eucalyptus and black wattle woodlots in western Kenya provided at least 10 million Kenyan shillings a year in farm income. In the Kakamega district alone, the estimated benefit from the protective role of farm trees in terms of higher crop production is some 30 million Kenyan shillings a year.

An important economic side-effect of this role of farm trees in increasing crop yields is the possible benefit in terms of reduced deforestation. If trees can help intensify crop (or livestock) production they may at the same time reduce the need to clear woodlands and other tree resources to make new farming land. This secondary benefit – and its further benefits in terms of the environmental damage which can be caused by deforestation – does not appear to have been considered in economic appraisals of farm forestry.

NEW THINKING ON TREE ECONOMICS

One thing that is obvious from this brief review is that the economics of farm forestry, with its multiple outputs, are not simple. In this section we argue that even if one restricts oneself to wood production from trees, and ignores their many other service functions, the economics are still not simple. In particular we suggest that, even on this limited perspective, conventional forestry thinking and economic appraisals must often be radically revised when dealing with rural tree growing and its rationale for local people – and that, by implication, many analyses of rural tree growing have as a result gone badly wrong.

Farm Tree Versus Forest Growth

John Spears, until recently the chief forestry adviser to the World Bank, has pointed out that forestry planners often think quite wrongly of rural tree growing in terms of area of forest rather than individual trees.[48] Forestry planners depend upon copious sets of data about tree growth rates for different species under various conditions of soil, rainfall and management, but these are mostly derived from experience of natural forests and plantations, where trees are closely spaced and are left to mature. Farm trees, on the other hand, are typically not packed together and therefore absorb much more sunlight. They are also frequently lopped for fodder or fuel, which can remove older and less vigorous growth.

As a result, Spears suggests that free-standing farm trees may grow at anything from three to six times faster than the same trees under the same conditions in a plantation. Recent but unconfirmed studies at the Oxford Forestry Institute suggest that heavily lopped farm trees in the Sahel can grow as much as ten times faster than the same species when left to mature in plantations.[49]

There is an urgent need to improve on these estimates with more and better field data. But clearly, if they are anything like correct, the need for rural tree planting to meet particular needs such as fuelwood may often be greatly overestimated and farmers wrongly discouraged from taking it up. If farmers each have to plant only forty trees, say, to meet their fuelwood needs, rather than 200 or more as anticipated by local plantation yields, they are very much more likely to be interested – especially if they face the usual constraints of lack of labour or land or cash for the materials.

Timber or Twigs?

Foresters from industrialized countries are used to thinking of trees as a means of producing timber, in which case long growing periods to increase stem diameters are an advantage. Most Third World farmers, on the other hand, are more interested in slimmer or smaller pieces of wood, particularly for construction poles or fuelwood. Their time frame and economic rationale for tree growing is therefore quite different.

This crucial point is illustrated by Figure 1.1, which shows how the net value of trees varies with stem diameter (or, roughly speaking, time), comparing the widely divergent cases of industrial forestry as exemplified by Europe and forestry for local needs in Africa.[50] Net revenues are derived from the wood value less harvesting costs. Revenue is

Figure 1.1: Wood Value as a Function of Stem Diameter: European and African Forestry Compared

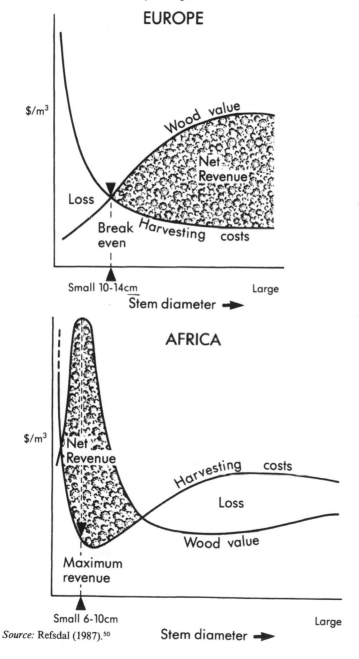

maximized in the European case when stem diameter is large, since this gives economies of scale in the processing of wood for industrial timber and wood pulp. Harvesting costs also fall with stem diameter owing to economies of scale with mechanization: it is cheaper to fell and cut a few large trees than many smaller ones of the same total volume.

In contrast, net revenues from smallholder forestry in Africa are maximized when trees are harvested at a young age when they have small stem diameters. This is because the major end-uses are slender construction poles, fuelwood and fodder. Even timber and charcoal making require trees of little more than around eight years' growth. At the same time, the labour requirements (and hence costs) of harvesting trees by hand and machete increases substantially with stem diameter.

The fact that most farmers cannot wait for long to reap the benefits of their investments must also be factored into this kind of comparison. This point has been forcefully made by an economic comparison in Kenya between woodlots which produce most of their cash benefits only after they have matured for ten years and a coppicing system which produces "fuelsticks" practically every year from the second year after planting. While the woodlots produced slightly more wood and cash revenues overall, the vast majority of farmers would want quicker returns and would greatly prefer the "fuelstick" approach. The details of this comparison, which had a major impact on woodfuel planning in Kenya, are presented in Rural Case 14.

Mixed Strategies for Rural Trees

Berry van Gelder has recently taken these ideas further by suggesting a sophisticated scheme of mixed "production cycles" for managing trees and shrubs in farming systems, based on evidence from the Kisii and Kakamega districts of Kenya.[51] The proposed management improvements build on farmers' local knowledge and existing agroforestry practices, and are aimed at satisfying various local needs for wood products.

The three most important types of tree stocks on farms in these areas are woodlots, trees in cropland, and hedges. At present only woodlots have a clearly defined cash function. At various intervals depending on species and local conditions they are clear felled and sold, usually for poles or building timber. Rotation periods are, in general, for eucalyptus, four to five years in Kakamega, and five to twenty years in Kisü; and for black wattle (*Acacia mearnsii*), eight to twelve years.

The existing management of hedges and trees in cropland follows a much more *ad hoc* production pattern. Trees in cropland may be pollarded or pruned at irregular intervals, with most of the wood going for use on the farm for building, farm tools or fuelwood. Trees in hedges grow fast and are often cut back every one or two years, but wood production *per se* is not the primary objective.

These systems could be improved by emphasizing both the multiple functions of trees and local production objectives. A method of designing such improvements which could form the basis of extension programmes is illustrated in Figure 1.2, which looks at ways of managing the wood stock in terms of several "production cycles" which are in turn based on three closely related aspects of tree management: an appropriate time frame (or tree rotation period), the planting density, and the size of wood output.

Figure 1.2: Nested Rotation Cycles for the Improved Management of Trees and Harvesting of Wood Products on Farms

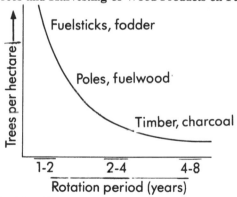

Source: van Gelder (1987).[51]

For example, if the farmer wants to produce poles, but the existing planting density is too high, a thinning plan can be proposed to improve production. If the trees are already overgrown, the wood could be used to produce charcoal or timber. The production of existing trees in cropland can be enhanced by improving pollarding systems. For example, pollarding at the start of the rains reduces the shading of crops as well as making more effective use of rainfall.

The KWDP is now trying out some of these ideas with local farmers in the Kegoye area of southern Kakamega. One technique is to stagger the ages of trees grown within a single farm woodlot, so that there are

Table 1.4: Findings of the KWDP Biomass Inventories in Three Districts of Kenya

1. Relationship between population density and wood stocks on farms in Kakamega district

rural households per km²	total woody biomass (t/km²)	per cent planted trees	per cent planted stands	per cent natural woody biomass
<50	420	10.3	24.3	65.4
51–100	472	10.9	34.1	56.0
100–150	1,152	12.6	40.8	46.6
>150	1,296	16.7	57.5	25.8

2. Agroforestry practices in the three districts

Percentage of farmers who had in the previous year:	Kakamega	Kisii	Murang'a
Planted tree seedlings	72	50	79
from own tree nursery	38	20	21
from "official" nursery	25	35	29
from neighbour's nursery	49	33	28
Collected "wildings"	45	28	71
Directly seeded	26	13	41

3. On-farm wood stock profiles for the three districts

Percentage of farms with:	Kakamega	Kisii	Murang'a
Trees in croplands	98	91	97
Trees in compounds	97	83	93
Woodlots	65	75	26
Windrows	–	53	43
Hedges	96	100	97
Bush	47	19	26
Natural woodland	–	–	10

Source: Chavangi and Ngugi (1987).[52]

always trees at different stages of growth which can be used for different purposes. The aim is to permit the continuous harvesting of wood products as and when they are required, unlike the present practice which is to harvest all the trees at one time. As for the choice of tree species, particular attention is paid to their multipurpose value and coppicing abilities, rather than the usual conventional forestry measures of overall tree growth rates.

Social Contexts of Tree Economics

The final example comes from related work in the same district and is a forceful reminder that the economics of rural tree management should never be appraised without considering the broader context of local people's livelihood systems.

The high potential farming areas of Kenya and other African countries often have surprisingly high standing stocks of woody vegetation, despite their high population densities. This fact was strongly underlined by biomass inventories conducted by the KWDP in the Kakamega, Kisii and Murang'a districts of Kenya, summarized in Table 1.4. Although it was known that there had been a rich tradition of tree culture in these areas for a long time, the surprise finding of the survey was not just that there is a great deal of wood around, but that the volume of farm trees actually increases with population density. Furthermore, as population density increases (and farm sizes fall) so does the proportion of trees and other kinds of woody biomass which have been deliberately planted (see top of Table 1.4). This finding at first appears almost counter-intuitive. Why, in particular, should farm plots which are so small as to be virtually sub-economic have up to a quarter of their area under trees?

Preliminary research by Peter Dewees suggests four linked explanations.[49] First, tree management and the sale of products to local markets give lower returns per hectare than cash crops (notably coffee) but higher returns per unit of labour. Second, small farm households often face severe labour shortages, since a high proportion of members are forced to earn cash incomes off the farm. Third, tree management has lower start-up costs than cash cropping, so it is more easily affordable by the smallest (and poorest) farmers. Fourth, while many of the latter could be interested in getting into cash crops rather than trees, adequate credit facilities to help them make the first investment are rarely available to them. Indeed, to get around this barrier some small farmers have sold trees to raise the funds to do a little cash cropping when crop prices have been particularly high.

This example brings us back full circle to where we began this chapter, with the complexity and diversity of African farming systems and the objectives which rural people apply to them. No amount of desk-bound theorizing about the economics of rural tree growing is any substitute for discovering the realities on the ground and basing one's choices of actions on them.

Chapter 2

Forestry for Land Management

In this chapter we review some of the most important technical options for integrating trees into farming systems, whether these are primarily agricultural and based on cropping, or pastoral and dependent on livestock, or some mixture of both. The emphasis is on trees outside the forests.

Virtually all of these techniques could be described as "agroforestry", a current buzz-word of development forestry, natural resources management and rural development. Part of its definition by ICRAF states that "all agroforestry systems embody the principle of deliberately using the special productive and protective features of woody plants to increase, sustain and diversify total output from the land". This definition is so broad that "agroforestry" can all too easily become a term which simply includes too much to be analytically meaningful. In practice, it is a plethora of rural tree growing options, encompassing a multitude of species, tree crop configurations and management practices – all in a great diversity of human and ecological settings.

Given this diversity, agroforestry clearly also varies greatly in its potential. On the one hand, in many places it shows considerable promise for addressing some of the most pressing problems facing land users: increasing food crop production and opportunities for earning cash incomes, maintaining soil fertility, reducing erosion, improving the micro-climate and producing fuelwood. On the other hand, it is by no means free of problems or the panacea it is sometimes made out to be.

In dry regions, for instance, there are agroforestry systems which can deliver extraordinarily positive and multiple benefits – but the trees can take fifteen years or more to grow. Farmers and herders may have no interest in waiting that long. Where land, labour or cash is scarce, farmers may simply be unable to afford to risk experimenting with agroforestry systems which could be a great boon if they work but could leave them worse off if they do not.[1] Nor are trees the only way to meet objectives: if increasing soil fertility is the concern, beans, grass and clover may be a cheaper and easier approach than planting leguminous trees.

Other warnings about trees in farming systems could be made: some argue that they make the rich richer and the poor poorer; they reduce the yields of certain crops; they harbour crop pests and evil spirits; and they can lower the water table. "Agroforestry" does not succeed everywhere, and one needs to be particularly wary of the presumption that it works well for resource-poor farmers. There are many cases of agroforestry and social forestry innovations or external projects that result in the expropriation of resources from communal use, or use by poor people, or by specific groups, particularly women.[2]

Table 2.1 summarizes the major biological and economic advantages and disadvantages of agroforestry. It suggests that agroforestry should not be regarded as a technical fix unless it is seen to be worthwhile by a farmer or pastoralist – or a community – or unless readily available modifications or incentives can make it so.

The important question, then, is *how* trees may be integrated in rural livelihood systems to address selected problems. This is particularly true of "agroforestry". The ways in which agroforestry systems are managed can make all the difference between "success" and "failure" in satisfying the basic needs of rural people. Trees certainly can have great potential in rural livelihood systems, but may not be the main concern of local people. The challenge is to facilitate a *process* of rural development – often but not necessarily involving trees – which puts people's priorities first.

The technical rural forestry options (and their economics) are nevertheless important. In many places, the standing stock of wood resources is diminishing, and access to wood for particular groups of people is a problem which may only be resolved by increasing the supply of trees. There will be little chance of achieving this objective if the available forestry options are not up to the job and so (as we noted in Chapter 1) the technical options must be good ones.

Throughout this chapter, where we refer to case studies the main emphasis is on their technical components rather than on institutional aspects or on different intervention approaches. These other issues are dealt with in Chapter 4. The technical options we discuss range from "pure" farmer-led initiatives such as private woodlots or traditional agroforestry in crop lands, through to outsider-led initiatives such as planned windbreaks or water-harvesting microcatchments for trees.

Table 2.1: Agroforestry: Advantages and Disadvantages

advantages	disadvantages

Biological:

• Better use of ecological space; captures more solar energy	• Competition for light between trees and other plants may lower crop yields
• Temperature extremes are reduced	
• More biomass returns to the soil	• Competition for space between trees and other plants may handicap both
• Recycling of nutrients is more efficient	
• Trees improve soil structure by producing stable aggregates and by avoiding hard "pans"	• Trees compete for nutrients, store them in branches and stems, and so make them inaccessible to crops
• Fewer weeds because less light reaches the ground and there is the possible suppressive effect of leaf litter mulch	• Loss of nutrients when wood, fruit, seeds, etc. are harvested and "exported" from the area
• Leaf mulch reduces water evaporation from soil, adds organic matter and reduces tillage needs	• Trees keep part of rainfall in their crowns; stemflow can adversely redistribute rainfall
• Most leguminous trees fix nitrogen by the action of specialized bacteria in the plant roots	• Greater diversity of fauna owing to a larger number of ecological niches; some will be crop pests
• Erosion is prevented up to a point by the binding effect of tree roots	
• Greater diversity of fauna owing to a larger number of ecological niches; some will be predators of harmful insects or rodents	

Economic and Social:

• Direct economic benefits in form of firewood, posts, poles, timber, fruit, fodder, etc. (although not all at once)	• Yields of crops per unit area may be lower than for monocultures
• Where commercial markets exist, trees constitute "standing capital" to pay for emergencies	• Even if combined value of trees and agricultural crops is higher, it may take several years for the trees to acquire economic value
• Crop diversity reduces risks of irregular rainfall, pest outbreaks, market fluctuations, uncertain supply of external inputs	• Likely to be more labour-intensive than growing either trees or agricultural crops separately
• Greater benefits from crops may offset investments required to establish trees	• Time-lag from planting to economic benefits of trees may be longer than people can afford by comparison with other cash crops
• Trees usually reduce weeding costs	
• Greater flexibility to spread work loads during the year	

Source: Budowski (1984).[3]

The socio-economic and policy aspects of various case studies are mentioned here only briefly as they are dealt with in other chapters.

DEFINITIONS AND TYPES OF AGROFORESTRY

For the sake of clarity, it may be useful to define some terms which are used for rural forestry and describe briefly the major systems of agroforestry.

Community Forestry (Social Forestry)

Community forestry programmes are based on the use of public or communal lands for tree growing. Though generally designed to meet community needs, programmes can involve very different levels of community involvement and participation. Forestry departments, in particular, can range from being a catalyst for community action to managing and implementing the planting, harvesting and disposal of the crop.[4]

Farm Forestry

Farm forestry is the term usually applied to programmes which aim to encourage commercial tree growing by individual farmers on their own private land. In these programmes, trees are regarded as a cash crop, and farmers are provided with assistance in growing them. This may include technical help, free or subsidized seedlings, loans and various market support measures.[4]

Agroforestry

Agroforestry is the collective name for all land use systems and practices in which woody perennials are deliberately grown on the same land management unit as crops and/or animals. This can be either in some form of spatial arrangement or in a temporal sequence. To qualify as agroforestry, a given land use system or practice must permit significant economic and ecological interactions between the woody and non-woody components.[5]

This comprehensive definition of agroforestry encompasses virtually all of the technical options considered in this chapter. Table 2.2 gives a

Table 2.2: Classification of Agroforestry Systems

Classifications may take one, or a combination, of four forms:

1. Ecological region (in tropics)
    - humid/sub-humid lowlands
    - arid and semi-arid lands
    - highlands

2. Structure (nature of components)
    - crops and trees (agrisilvicultural)
    - trees and pasture/animals (silvopastoral)
    - crops and pasture/animals and trees (agrosilvopastoral)

2a. Structure (arrangement of components)
    - spatial (mixed dense: e.g. home gardens; mixed sparse: e.g. most silvopastoral systems; strip; or boundary)
    - temporal (coincident, concomitant, overlapping, sequential, separate, interpolated)

3. Function, mainly of woody perennials
    - productive (e.g. food, fodder, fuelwood)
    - protective (e.g. windbreak, shelterbelt, soil conservation, soil improvement, moisture conservation, shade for crops, animals and people)

4. Socio-economic basis
    - level of technological inputs
    - intensity of management
    - commercial goals (or otherwise) or cost/benefit relations (commercial, intermediate or subsistence)

Source: Nair (1987).[6]

broad classificatory framework for agroforestry systems, arising from the global inventory of existing agroforestry systems and practices in the tropics and subtropics recently completed by ICRAF. The fundamental basis for classification is in terms of stucture: that is, in the components of the system and the way they are arranged with each other. All agroforestry systems comprise some combination of at least two of the three main components: woody perennials, pasture and animals, and agricultural crops. A classification of agroforestry systems on the basis of their components is presented in Figure 2.1.

The physical setting of agroforestry is of course immensely diverse and so one finds considerable differences in the types of agroforestry and mixtures of its components. The following brief and generalized summary of what agroforestry systems are found where in Africa is taken from Nair.[6]

Figure 2.1: Categorization of Agroforestry Systems on the Basis of their Components

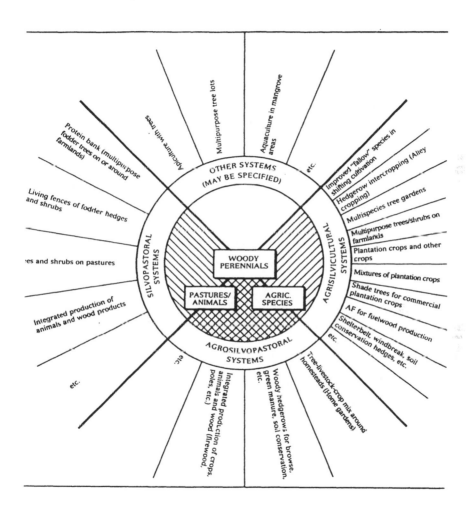

Source: Nair (1985).[7]

Agroforestry Systems in Arid and Semi-arid Lands

These extend over the savanna and Sudano-Sahelian zones of Africa. Drought is a hazard throughout much of this zone and pastoralism is the major livelihood strategy. The main agroforestry systems vary considerably depending on climate and human population pressure; home gardens and multilayer tree gardens, for instance, are common in the wetter areas of semi-arid lands with high population pressure. The predominant systems in the zone are:

- various forms of silvopastoral systems
- windbreaks and shelterbelts
- multipurpose tree growing on crop lands, notably systems based on *Acacia albida*.

Agroforestry potentials in this zone include anti-desertification measures and fodder provision as well as fuelwood supply.

Agroforestry Systems in the Tropical Highlands

Such systems are found in many parts of East and Central Africa. Altitudes exceed 1,800 metres in substantial areas of the Kenyan and Ethiopian highlands. The highland tropics with significant agroforestry potential are humid or sub-humid. Sloping land and marked relief, combined with unpredictable and frequently heavy rainfall, make soil erosion an issue of major concern. The relatively low annual temperatures limit the growth of certain lowland tropical species. The main agroforestry systems in these areas are:

- production systems involving plantation crops such as coffee and tea under both commercial and smallholder management
- use of woody perennials in soil conservation and soil fertility maintenance
- improved fallows
- silvopastoral combinations.

Agroforestry Systems in the Lowland and Sub-Humid Tropics

These are the most important ecological regions of Africa in terms of the total human population they support, and in the area and diversity of agroforestry and other land management systems. Agro-climatic conditions favour the rapid growth of a large number of plant species and a great variety of tree crop associations can be found in areas with high human population. Common agroforestry systems in these areas include:

- various forms of home gardens
- plantation crop combinations (such as coffee and banana with food crops)
- multilayer tree gardens
- multipurpose tree woodlots
- improved fallow in shifting cultivation systems (in areas of low population density)
- alley cropping and alley farming (i.e. with livestock).

The analysis of ICRAF's agroforestry systems information base has shown that the existence or adoption of a particular agroforestry system in a given area is determined largely by the *ecological potential* of the area. Within this framework, specific socio-cultural and economic factors determine the complexity of the system and the intensity of its management.

For a given set of agro-climatic conditions, the socio-economic determinants of local agroforestry systems include factors such as *human population pressure*, availability of *labour and other factors of production*, and proximity and accessibility to *markets*. But most important of all in shaping the impact of these factors are the differently ranked needs and priorities of local people, and how tree growing fits – or does not fit – into their production strategies as a whole.

OPTIONS FOR AGRICULTURAL AREAS

Enriched Fallows

The shortening of fallow periods in the traditional shifting cultivation cycle continues to be one of the main factors in land degradation in the tropics, while these extensive systems can support only low population densities. Agricultural intensification to remedy this can involve the usual, relatively high-cost technical packages typified by "seeds, water and fertilizer" but can also involve forestry approaches to enrich fallows. One example of this approach was presented in Chapter 1: the *chitemene* bush fallow system in north-east Zambia is being intensified in places by farmer-led experimentation with soil mounding and composting (see Rural Case 7). ICRAF's Agroforestry Diagnosis and Design methodology has been applied to this system to identify further ways of increasing its productivity.

Fallows may be enriched in two ways: either economically or biologically. Economic enrichment involves growing trees for cash or subsistence purposes. An indigenous example of this is the growing of gum arabic in the Sahel, particularly in the Sudan.[8] Biological enrichment, on the other hand, is aimed primarily at increasing soil fertility, for example using nitrogen-fixing leguminous tree species such as *Leucaena leucocephela*.

In Kakamega district of Kenya, for example, a common traditional practice in fallow enrichment is to scatter seeds of the leguminous *Sesbania* tree on fallow land. The growth of woody vegetation in this high potential area is rapid, and quite apart from improving soil fertility, the trees provide fuelwood when the time comes to clear the land to make way for agricultural crops.

Multi-Storey Home Gardens

In certain relatively humid areas, some indigenous agroforestry systems make intensive and efficient use of limited space. A good example is the system of *Chagga* home gardens, covering some 120,000 hectares on the southern and eastern slopes of Mount Kilimanjaro in Tanzania. The plots average only 0.68 hectares and yet they meet between a quarter and a third of farmers' fuelwood needs and often all of their livestock fodder requirements. At least fifty-three species are cultivated, 90 per cent of which produce fuelwood and 10 per cent fodder. Roughly a quarter of all cultivated species produce poles and timber, some 30 per cent produce medicines, and 19 per cent serve a range of other puposes such as providing mulch, pesticides, fruit and shade.[9]

Other examples of richly diverse multi-storey home gardens can be seen in south-west Kenya, such as the two tiny farms of Ainea Mundanya and Joseph Mogaka which were described briefly in Chapter 1.

Alley Cropping

In contrast to the practice of rotating fallows over time, alley cropping achieves "continuous fallow" in space through the simultaneous association of trees and field crops. It has been defined as a "zonal" system of agroforestry[10] in which field crops are planted in the alleys between hedgerows of nutrient-cycling trees or shrubs. Fodder and fuelwood might be taken as by-products, but the basic aim of the trees and shrubs is to provide a soil-enhancing service to arable farming. The FAO has described such systems as "probably the most versatile, effective and widely adoptable of recent innovations in conservation farming".[11]

Alley cropping was pioneered at the International Institute for Tropical Agriculture (IITA) at Ibadan in Nigeria, using *Leucaena leucocephela* hedgerows with maize growing in the alleys. Although Ibadan lies in the sub-humid zone, it appears that alley cropping may have potentials for a wider ecological range. Indeed, it appears to be not only technically possible in semi-arid areas but is even working well in such a zone at the Machakos research station in Kenya, run by ICRAF.

In livestock-keeping areas where fodder for animals is an important basic need, a variant of alley cropping known as alley farming may be suitable. For example, as noted in Chapter 1 (and Rural Case 8), an alley-cropping system developed by Kenyan research scientists was modified by local farmers to become an alley farming system because they felt a greater need for more animal fodder than for better food crop yields.

An impression of the potential yield increments from alley cropping was given in Table 1.3 (p.46). The increments in maize yields in the trials and extension efforts reviewed in the table were in the range of 39–75 per cent on farms and up to 83 per cent on-station. The additional fuelwood yields in each of these cases were also substantial, ranging from 5 to 10 tonnes per hectare each year. This would typically be much more than an individual family's needs, so provided there are local markets the trees can provide additional income from fuelwood sales. However, one has to treat these examples with caution since there has not been enough time or a sufficiently wide experience of alley cropping to be sure whether such yields can be maintained or whether these are exceptional figures for particularly advantageous sites.

Alley cropping can also make good economic sense in terms of the productive use of farm labour. One study in Machakos district, Kenya, compared a conventional maize/bean cropping system with one in which *Leucaena leucocephela* alley hedgerows were interplanted with maize and beans.[12] While the alley cropping system showed higher costs than benefits in the first year, labour requirements decreased in subsequent years relative to revenues. This was mainly due to the reduction in labour demands for field preparation and weeding, since only half the total area was cultivated in the alley cropping system. Moreover, additional products such as fuelwood and fodder can be lopped from hedgerows at times when other demands on labour are low. However, if the tree species are to be managed for a range of outputs, then almost by definition farmers will have a lot to do. For example, if the trees are also to form a fodder bank, they may need to be cut back on most days.[13]

Boundary Tree Planting

Tree planting need not always compete with field crops for space, moisture and other resources: complementary and supplementary economic relationships between trees and crops abound in practice. A special opportunity for supplementary production through agroforestry is by planting trees in in-between places: along farm boundaries and internal borderlines as living fences; along paths, roads and watercourses; around houses and in public places; or in generally underutilized lands. Such plantings involve little or no opportunity cost except for labour.

The productive potential of this type of agroforestry has been demonstrated by analysis of aerial photographs of a watershed in a fairly densely settled farming community in Kathama sub-location of Machakos district, in Kenya. The study indicated that if existing linear features of the landscape – pathways, watercourses and boundaries – were fully utilized for the planting of appropriate trees and shrubs, some 50 per cent of fuelwood and 40 per cent of fodder requirements of local households could be met by these hedgerows, with very little competition with existing agricultural land uses.[14] Table 2.3 shows the estimates of the production potential which could be achieved were trees to be planted along all these linear features and boundaries.

Few data exist on the economics of hedgerow growing, except for certain well-documented alley cropping systems, as discussed above. However, hedgerows do appear to have some characteristics which can make them an attractive means of increasing wood stocks on farms. Many hedge species grow quickly, which makes them an attractive first step in (re-)integrating trees into farmland. If used as live fences, hedgerows also tend to be considerably cheaper than, for example, barbed-wire fencing. Some examples of comparative fencing costs are given in Table 2.4, and show that costs can be reduced, typically by a

Table 2.3: Production Potential Estimates of Linear Features and Boundaries for a Watershed in Machakos District, Kenya

linear feature of the landscape	roads & paths major	minor	gullies & channels major	minor	boundaries & internal borders existing	bench risers/rows	total
1. Length (m)	2,600	2,400	4,200	3,200	8,340	15,000	35,740
2. Width (m)	2	1	3	2	1	1	
3. Area (m²)	5,200	2,400	12,600	6,400	8,340	15,000	49,940 (5 ha)
4. Potential fuel production (kg/yr)	10,400	4,800	25,200	12,800	16,680	30,000	84,880–100,000
5. % watershed area	7%	3%	17%	9%	11%	20%	67%
6. Potential tree fodder production (kg/yr)	10,400	4,800	25,200	12,400	16,680	15,000	84,880 (84.9 t)

Characteristics of research site:
1. Total area of watershed: 106.5 ha.
2. Population density, Kathama sub-location. 172 persons/km².
3. The watershed represents approximately 7 per cent of Kathama sub-location.

Assumptions:
1. Production estimates assume 1 lopped tree per square metre producing 2 kg dry matter leaf and 2 kg dry wood per tree.
2. 66 per cent of area in grass strips, 33 per cent in fodder trees (production estimates for grass fodder not included here).
3. Assumes only 500 metres of bench risers or rows per farm for 30 farms.

Source: Rocheleau and van den Hoek (1984).[14]

factor of three. The cost savings depend on the methods used in seedling production, however, and costs may still be prohibitive to farmers unless the seedlings are raised in mini-nurseries on individual farms, or by direct seeding by farmers. This point is illustrated in Rural Case 15 from the Koro Village Afforestation Project in Mali.

Nevertheless, while live fences and other hedgerows can produce a wide variety of outputs, they can best serve only one main purpose at a time. For example, most leaves are produced for fodder when the hedge

Table 2.4: Comparative Costs of Barbed-Wire and Live Fences

country	barbed-wire fencing US$/m²	live fencing US$/m²	ratio of cost difference
Djibouti/Somalia	0.6	0.2	3
Kenya	0.5	0.3	1.7
Senegal (enclosure of *Acacia Senegal* plantation)	450/ha	150/ha	3

Source: Sepasat (1986),[15] quoted in Kuchelmeister (1987).[16]

is coppiced frequently, and so little seed is produced. If both seed and leaves are desired, the production of both will be lower than if the hedge were managed for the regular harvest of just one of these products. Seed production may not be a constraint, however, as seeds can usually be gathered from species growing outside the farm. In general, though, project design for multi-purpose agroforestry must inevitably involve some compromises of this sort.[16]

Boundary tree planting projects may also run into disputes over land tenure or rights to trees.[10] These issues are addressed in Chapter 3, but as a case in point, tree planting on apparently "underutilized" lands might conflict with the rights of those who currently graze animals on or gather from those lands. It is not a straightforward matter to devise equitable and workable controls for the use of multi-purpose trees and shrubs on common property lands. One promising approach which has been suggested is to partition planting responsibilities and exclusive harvesting rights among individual members of a work group, organized and operating at community level.[17] Such patterns of organization and of community sanctions may build on structures relating to grazing and water control, where these exist, but a degree of social cohesion which is rare among communities who occupy lands in marginal areas is also required.

In the Kathama Agroforestry Project, Machakos district of Kenya, for instance, it was found that nursery work needed to be organized as a group activity near a permanent source of water on the farm of one group member. The local capacity for such forms of social organization had already been demonstrated by the existence of agricultural self-help community groups in the area (see Rural Case 16).

Community Woodlots

The history of community woodlots in sub-Saharan Africa has been a chequered one. Their proliferation in some other regions, notably South Asia, has been made possible by marketing systems which allow villagers to earn more by growing trees than by growing crops. This rarely seems to be the case in Africa at present, although rising prices and the greater reach of commercial markets for tree products may alter this (see Chapter 6).

Major constraints to community woodlots in Africa relate to the heterogeneity of communities and the associated issue of land and tree tenure, as discussed in Chapter 3. One of the best-documented examples of how these constraints inhibit "social" forestry is provided by Margaret Skutsch's study of the "energy woodlots" programme in Tanzania, which is described in Rural Case 17.

In the context of the Village Afforestation Programme pursued by the government of Tanzania, Skutsch sampled eighteen villages to examine the constraints to communal tree planting, or "why people don't plant trees".[18] Some villages started woodlots while others did not, some woodlots failed once started, and planting often stopped in woodlots after their establishment.

In this case the main constraints preventing some villages from starting woodlots were the extra demands on agricultural labour time, the loss of private land for communal use, and low participation as a result of factionalism within the community and a lack of confidence in village governments. Other villages saw their woodlots fail because they made little impact on people's felt needs. The overemphasis on planting trees – at the expense of their sustainable management – underlies this failure, and itself has structural causes at a wider institutional level. The understanding of the constraints operating in this case may have more general application to similar tree-planting experiences elsewhere.

The slowness to adopt community forestry in this Tanzanian programme cannot be attributed to the lack of either an awareness of the "need" for trees or of technical knowledge. All villages in Skutsch's sample expressed a desire to plant trees and knew of the Tanzanian government's policy on village afforestation. Every village recognized that the supply of wood products from local natural woodlands was falling, while many households had already started to plant trees for themselves, mostly around their homes. But this is an insufficient basis for the successful establishment and management of *community* woodlots.

Private Woodlots

The constraints to successful community forestry identified in this Tanzanian study and elsewhere have led some to conclude that family woodlots – or true farm forestry – offer more tangible incentives for the participation of local people.[19] The Second World Bank Forestry Project for Niger ruled out community woodlots for this reason, as did the CARE-Uganda Village Agroforestry Project.[20]

Highly successful examples of private woodlots, based on production for the market, can be found in the high potential agricultural areas of

Kenya.[21] Conditions here are, of course, very different from those in the villages sampled by Skutsch for the study quoted above. In Chapter 1 we discussed the specific conditions of this Kenyan case, where the role of local markets, labour constraints, access to credit and other factors are all major influences (see Rural Cases 9 and 14). Generally speaking, it is more profitable to manage private woodlots for the production and sale of construction poles rather than woodfuels, which are often the (saleable) "rubbish" – or loppings, etc. – left over from pole production.

OPTIONS FOR LIVESTOCK-KEEPING AREAS

Biological Means of Erosion Control

Some of the technical options we consider in this section, particularly those in the drier livestock-keeping areas, serve not only to increase the production of woody vegetation but also to control soil erosion. Encouraging the natural regeneration of vegetation by livestock exclosures, for example, is a biological means of erosion control which in Africa has usually proved to be both more successful at controlling erosion, and more appropriate socially and ecologically, than outsider-led "engineering" methods such as the construction of terraces. This fact was emphasized in the World Bank/FAO Ethiopian Highlands Reclamation Study, which outlined a conservation-based development strategy, including agroforestry and improved livestock management, with erosion control as one of several objectives. Some highly successful strategies in runoff farming in Africa have used contour bunds which act as a part-"engineering" and, through their impact on the productivity of vegetation, a part-biological means of erosion control (see, for example, Rural Case 10).

Trees in Agropastoral Systems

The limited soil moisture in dry lands poses problems for crop husbandry which trees can help to overcome. One of the keys to coping successfully with this problem is to maintain adequate levels of soil organic matter and nitrogen. However, conventional green manuring and mulch farming are limited in these regions by the low production potential of herbaceous biomass (e.g. grasses) and by its competition with food crops for water, land and labour. Multi-purpose trees offer several advantages over herbaceous biomass in this respect:

1. they are generally more drought-tolerant
2. they are a valuable source of fodder during the dry season (particularly the pod-producing species) when herbaceous fodder is scarce
3. they can grow along boundaries on the farm or in association with crops without replacing them
4. they can offer micro-climatic benefits if appropriately spaced and placed with respect to crops
5. they can produce fuelwood, food, building materials, etc. while still performing their service roles on the farm
6. they are a form of standing capital and in certain cases can serve as convertible savings for meeting emergency needs, including "famine foods".

Trees can strengthen the role of livestock on the farm by enhancing the fodder-manure linkage. The growing of additional high quality dry season fodder on the farm and the use of living fences to control livestock access could potentially go a long way towards relieving the main sources of tenure conflict between farmers and pastoralists, or between the two land use systems where farmers are also pastoralists and vice versa. The key to securing this innovative type of arrangement seems to be to get the pastoralists to plant trees, rather than to expect farmers to plant in such a way as to satisfy pastoral requirements as well. The main constraint in realizing these benefits is the need to restrict livestock access itself, through social sanctions if possible, during the establishment phase of the trees.[10]

Trees in Cropland

Traditionally most farmers in Africa south of the Sahara cultivate their crops and graze their livestock under and among trees. One of the most outstanding examples is the association of *Acacia albida* with dryland grain crops in the Sahel.[22] Indeed, *Acacia albida* has been called a "miracle" tree for this and other reasons. It is the fastest growing of all savanna trees. It is relatively drought-resistant. It has palatable seed pods for animal fodder. Its edible leaves contain about 15 per cent protein. Its hard wood has many uses and its bark contains tannin for curing hides. And by pumping nutrients from the soil and tending to improve soil water retention under its canopy, it can greatly increase crop yields.

As noted in Chapter 1, the yields of crops grown in proximity to trees in these dry lands may be as much as double those of crops grown in the open. One study on the infertile sandy soils of Senegal's Groundnut Basin found that groundnuts and millet growing directly under *Acacia albida* foliage yielded 900 (±200) kilograms per hectare compared with yields of 500 (±200) kilograms in the open (see Table 1.3, p.46).[23]

Much of the increase in yields can be attributed to nitrogen fixation, microclimatic benefits and the peculiar "reverse phenology" of the tree, which leafs in the dry season and drops its leaves at the beginning of the rainy season, thus fitting in well to the cropping cycle. Another factor is the concentration of manure near the trees from the livestock that gather under their shade in the dry season to consume the nutritious pods produced at that time.[10] Felker has suggested that the presence of *Acacia albida* on farms could increase the land carrying capacity from 10–20 to 40–50 persons per square kilometre and enable farmers to lead a more permanently sedentary lifestyle by eliminating the need for fallow periods.[23]

However, a major constraint to such systems in drier regions is that the trees can take fifteen years or more to establish. If farmers are forced by the pressures of survival to take short-term horizons – or even to expect to move on every five years or so after depleting the soil by cropping or grazing out the local grasslands – they will have little interest in adopting the systems without substantial assistance from outside.

Another widely encountered traditional agroforestry strategy in semi-arid cropland is the growing of grains, pulses and cotton under an open canopy of shea-butter trees (*Butyrospermum paradoxum*) together with *Parkia biglobosa* trees. Many other tree crop combinations exist besides these examples. To outsiders such farming practices, with scattered trees standing in cropland, appear "messy", primitive and inefficient. However, they are usually the outcome of repeated, well-organized, selective thinning carried out over long periods of time, favouring those species which do not compete severely with crops for light and water and which even permit significant crop-yield increases and other environmental benefits valued by the farmers.[24]

Windbreaks and Shelterbelts

Windbreaks or shelterbelts are often proposed by outsiders as a way of providing a range of environmental benefits in dry areas where the removal of the natural woody vegetation is putting the land at risk, particularly from wind erosion. They are, however, models transferred from temperate regions, and in semi-arid Africa have had mixed results.

A shelterbelt project in Maroua region of Cameroon, although modest in size, has been in existence for about a quarter of a century. Widely spaced belts of *Cassia siamea* trees have been planted across a lowland plain, but they are dense and form almost complete obstacles to air movement. While they are undoubtedly preventing soil erosion by the wind, their overall effect on the farming system is less clear-cut and not altogether beneficial. In fact, Gunnar Poulsen observed that towards the end of the cropping season, on a 30-metre-wide strip on either side of the tree rows, the sorghum and cotton crops had almost totally failed.[24]

But there are also cases to be found where windbreaks, although taking up arable land, can still permit significant crop-yield increases. Project experience indicates that careful management of the windbreak trees is the key to their success in providing environmental benefits, while not jeopardizing farmers' production needs. Shelterbelts which are too compact, for example, may be opened up by thinning or pruning individual trees. Water absorption or over-shading by the trees may be reduced to desired levels by occasional pollarding.

The CARE Majjia Valley Windbreak scheme in Niger, which was mentioned in Chapter 1 (see Rural Case 12), is an example of a highly successful windbreak project, at least in its technical aspects. The project's aims of protecting and conserving the soil, increasing agricultural production and producing wood for use as fuel and poles, have largely been met; and local people are now planting new windbreaks on their own initiative. Through visits to the Majjia Valley, farmers elsewhere are adopting similar techniques (see Rural Case 15).

However, it must be said that the Majjia Valley is one of the most widely quoted forestry "success stories" of the Sahel and that – as yet – there are precious few, if any, comparable successes in other dryland areas.

Exclosure of Livestock

Most livestock-keeping areas are semi-arid and arid lands where pastoralism is the major economic activity and way of life. However, pastoralists have many uses for, and frequently an extensive knowledge of, local trees, which are often an important source of dry season fodder and browse, and well as other economic benefits. Rural Case 3 describes some of the ways the Turkana pastoralists of northern Kenya make use of trees, and how their importance has shaped patterns of ownership, access and control over those resources.

Encouraging the natural regeneration of woody vegetation is another important means of rehabilitating degraded arid and semi-arid lands where tree growing is more difficult than in wetter areas and is often virtually prevented by browsing livestock. Livestock exclosure is an attempt to get round the latter problem by fencing areas of overexploited vegetation to allow them to recover on their own.

In Tanzania, the HADO soil conservation project attempted to close off hillsides from grazing livestock but used rather authoritarian methods, including the enforced destocking of cattle (see Chapter 1 and Rural Case 13). Recently, however, cattle owners have generally become reconciled to the changes in their lifestyle and recognize the clear benefits derived from the regeneration of the local vegetation. In two years about half of the ground cover in the worst-affected Kondoa Eroded Area had returned after exclosure and destocking.

In other examples which were discussed in Chapter 1 (see Rural Cases 1–3), greater success with exclosure was achieved from the start either by securing the consent of local people, or, in some cases, by their taking the initiative themselves. These cases show that quite remarkable rates of regeneration can occur and that, if local people are aware of such tangible benefits, they can often innovate to provide the community controls and sanctions to ensure that exclosure is voluntarily enforced. However, even such a simple technique as exclosure can go wrong if not carefully managed. For example, one area in the Pokot district of Kenya was closed off due to armed cattle raiding and, without management, became impenetrable bush which grazers could not use.[25]

One pastoral situation in which the need for trees is particularly great is over-grazing around dry-season water sources. The community forestry project at Um Inderaba village in the northern Sudan is an example of a successful and innovative project, defined and managed by the villagers themselves, which has established a social mechanism for mitigating the ecological effects of herd concentrations around the local wadi and borehole well. This scheme, presented in Rural Case 11, involved exclosures as well as a windbreak and other forms of tree planting, and the protection of young trees by means of a village guard rota.

Rainwater Harvesting

This technique is not so much a forestry option as a vital prerequisite to tree growing in dry areas. The principle is to concentrate into the cultivated area runoff from a larger catchment area. In arid regions with annual rainfall as low as 100 millimetres, the ratio of catchment area to cultivated plot may be from 10:1 up to 30:1; in semi-arid areas with annual rainfall of around 500–600 millimetres, the ratio may typically be from 5:1 to 20:1.[26] Two means of concentrating rainfall in this way are the construction of earthen and/or stone contour bunds or ridges, and the construction of V-shaped microcatchments.

In many dry parts of Africa, farmers have developed a wide technical knowledge of runoff farming techniques and the use of contour barriers and small check-dams for erosion control. The successful development of stone lines for water harvesting in the Yatenga region of Burkina Faso, through a fascinating combination of indigenous knowledge and joint research with a donor agency (OXFAM), was discussed in Chapter 1 (see Rural Case 10). Crop yields have increased by around 10–90 per cent (see Table 1.3, p.46) and the technique is being taken up very rapidly all over the Yatenga Plateau, with thousands of hectares now treated. Although the initial interest was in raising crop yields rather than trees as the project originally intended, some farmers are beginning to plant trees along their contour bunds as well as encouraging the natural regeneration of trees and bushes where the barriers have stabilized the soil.

The use of microcatchments in arid areas was illustrated in Chapter 1 and Rural Case 3 with an example from the Turkana district of northern Kenya. On a trial site, microcatchments of different sizes have been tested with a range of exotic and indigenous tree species. Even outside the trial sites, in areas of rangeland which are completely unfenced, quite staggering growth of planted trees has been achieved, with the commitment of local herders. Recognizing their value, a pastoral people with traditions of "free range" migrant herding have started to protect the young trees purely by social agreement. That is a remarkable change.

A NOTE OF CAUTION

Following that last "success story", we end this chapter with a word of caution. Demonstrations of water harvesting funded by the World

Bank's Baringo Pilot Semi-Arid Area Project in Kenya showed enormous crop yield increases, of 2.3 to 3.4 times for sorghum and 3.5 to 7.7 times for cowpea. Yet local farmers did not take the practice up, partly because they could not spare the extra labour.

Surprising "adoption gaps" like this are common in sub-Saharan Africa, perhaps most particularly with tree growing in the drier lands where survival is often marginal and there is little of anything – cash, labour or risk-taking confidence – to spare. Biological productivity, stocking densities and crop yields are often so low that large incremental gains have to be made as a first condition of farmers' interest. Time horizons are necessarily short: quick returns are often vital for success. This is especially true for longer-term "fixed technologies" like trees, when one may have to move on to new grazing or cropping lands in a few years' time. The great variability of the climate as well as generally low rainfall is another major barrier to successful innovation. As one Australian farmer put it, summing up rather well the dilemma in all of these risky areas, "You can learn enough in one year to make a fool of yourself in the next."

As this and the previous chapter have shown, there are many agroforestry and other afforestation packages to suit a wide range of conditions and, in some places, these have brought such substantial gains to livelihood that local people are likely to sustain and replicate them. It is now time, however, to look at the many reasons why these technologies may not "work". In the next two chapters we look at the constraints on rural tree growing and how they might best be met by outside agencies.

Chapter 3

Constraints on Change

The diversity of landscapes across sub-Saharan Africa is matched at the local level by equally diverse and complex rural livelihood strategies. These in turn are based on the complex economics of household production and consumption, with their many constraints and competing demands for labour and cash, and on equally complex social arrangements regarding access to land and trees.

Projects and programmes which aim to increase the economic productivity of rural livelihood systems and improve access of the poorest to tree products must take these social dimensions of local diversity fully into account. Although they sometimes provide the key opportunities on which interventions can be based, more often than not they add up to critical constraints. At the very least, project success depends on recognizing these constraints; at best it means overcoming or reducing them.

In this chapter we look at the key constraints on rural livelihoods under three headings: household economics, tenure and rights to land and trees, and gender divisions within the household economy.

HOUSEHOLD ECONOMICS, LABOUR AND CONFLICTING NEEDS

Rural people always assess, informally, the relative costs and benefits of investing their time and energy in different tasks. There are usually explicit trade-offs between these demands on labour, and between expected benefits in the short term and those in the long term. One dimension of poverty is the frequent necessity to forgo long-term benefits in order to meet short-term needs. This is the vulnerability aspect of poverty: having to spend one's savings or liquidate one's assets rather than saving them "for a rainy day".

Resource-poor farmers, whose livelihoods are marginal both economically and ecologically, tend to be more concerned with the management of risk than with most other economic considerations. Often, this entails avoiding any change in the allocation of household resources (labour, capital, etc.) which, though it might improve the situation if it works to the householders' advantage, could leave them even worse off if it does not.[1] Rural tree growing may either increase or decrease economic risk, according to whether or not it is in competition with other household activities.

The availability of labour is bound up with the issues of "needs conflict".[2] The trade-offs between meeting various subsistence needs and earning cash to meet other needs, for example, make competing claims on household labour. It is important to focus on poverty and vulnerability, since inequalities between rural people – both in social and economic terms – are one important determinant of their relative abilities to meet basic needs.

Unless outsiders understand local rankings and prioritization of needs, they may fail to appreciate why people may not be interested in or are prevented from growing or managing trees. Many other more pressing needs, like food security, may come before tree-related needs.

The Rural Squeeze

There are crucial linkages between problems of labour availability, needs conflict, and destructive land-management strategies,[3] which Ben Wisner has called the "rural squeeze".[2] The precise components of the rural squeeze vary from place to place, but everywhere it hits resource-poor households hardest.

Two woodfuel-related examples from Kenya and Lesotho show how the processes of privatization, commoditization and marginalization combine to squeeze the coping abilities of poor rural households. In each case, while the effects in terms of access by the rural poor to woodfuel are superficially similar, the actual processes are quite different.

In Kenya, land privatization and the establishment of individual rights over resources have led to a decline in access to commons and their products, such as roofing thatch, "bush foods", fuelwood and free grazing. This puts increased pressure on family plots, since more needs have to be satisfied by them. At the same time, the commodification of previously free or bartered subsistence goods is increasing the family's need for cash to buy essentials, including fuelwood. To raise cash, males are increasingly leaving the farm for wage employment, thereby reducing the labour available for farm work. Putting all these processes together, families are increasingly marginalized: that is, the productivity of family land falls relative to increasing needs.[2,4]

In Lesotho, the primary cause of the rural squeeze is integration into the South African mine labour system, which has left many women as providers of their family's basic needs, including domestic fuel, in a highly commoditized rural economy. The crisis of access to woodfuel is suffered mainly by women – particularly widows and divorcees – who do not receive cash remittances from employed male relatives.[4]

Phil O'Keefe[5] has shown in a most graphic way how processes like this can connect land and natural resource degradation in a vicious spiral which works back through urban poverty to rural poverty and labour shortages: see Figure 3.1.

As O'Keefe puts it, rural poverty is the essential cause of environmental degradation in sub-Saharan Africa. Artificially low prices for farm produce, involving massive subsidies for urban consumers at the expense of rural people, are largely to blame. To increase incomes, young and adult men from rural areas have to seek employment elsewhere. Although South Africa is now a major employment magnet for all the "frontline" African states, this pattern was set long ago by the

Figure 3.1: The Cycle of Poverty and Land Degradation

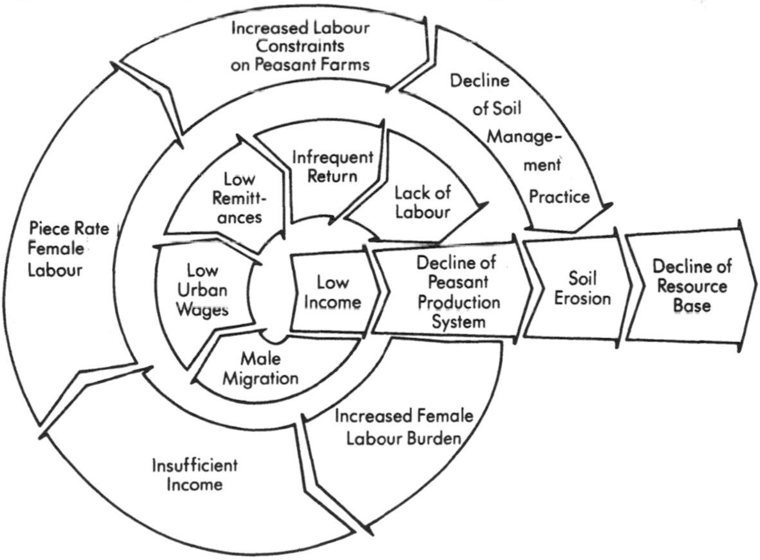

Source: Based on O'Keefe (1987).[5]

colonial powers which deliberately taxed subsistence rural producers in order to create cheap labour reserves for the urban mines and factories. But competition for urban jobs from this flood of rural migrants leads to or maintains low wages. Consequently, the migrant workers can afford neither to send back much cash nor return often to their rural families.

With a depleted labour force and little cash to invest in farming or to hire help, rural women bear such heavy work burdens that farm production declines. To offset this and buy basic essentials, women are forced to seek waged employment off the farm – thus deepening the vicious cycle of labour scarcity. Often this leads not just to lower production – for example, because planting is too late or weeding too hurried – but to a more permanent and irreparable failure to maintain soil quality and the environmental basis of agricultural potential. Alternatively, the family may have to sell land or other vital productive assets to raise cash or reduce labour burdens, increasing the grip of the rural squeeze and marginalizing them still further.

These processes place pressure on rural households, particularly the poor, to diversify their economic base. The most critical short-term need is often to increase cash income but, as we have just seen, this can involve difficult trade-offs with labour and other vital resources. As land and/or labour resources decline, one frequent result is the abandonment of multiple land management practices to meet food and other subsistence needs and a switch to producing the cash crops which government extension agents may be promoting.[2] Farmers become drawn into production for the market, which in time becomes more essential as their production of subsistence goods – including fuelwood and other products from trees – declines.

The need for cash can also divert labour and other investments from food production or positive resource management practices to other ways of producing income on the farm. Beer brewing, charcoal production and brick making are among the most common of these "non-agricultural" cash-earning activities – all of them extremely energy-intensive, thus adding to pressures on local wood resources. Although such non-agricultural sources of income are important to most farmers, they are especially important to the poorest. In Kenya for example, surveys in the mid 1970s showed that in the lowest-income group, 77 per cent of rural income came from such sources or employment off the farm, and 47 per cent for the next group up the income scale.[6]

These conflicts between needs and assets are also felt between households, as a result of inequalities in needs and access to resources. This was a major constraint, for example, to the success of the Forestry-in-Grazing-Lands project in the White Nile province of Sudan, which was organized by the Sudanese Forestry Department. In this dry area, the project aimed to plant trees to halt soil erosion and sand-dune encroachment; to increase dry season fodder for livestock; and to provide fuelwood and building timber. However, the conflict between the need for dry-season fodder for some and the need for cheap fuelwood for others became heightened with the increase in rural inequalities. The better-off, with large herds and with sufficient cash to satisfy fuelwood needs on the market, would benefit most from the improved (tree) fodder supply. The poor, with few, if any, animals but a continuous need for cheap fuelwood, would be forced to cut the young trees for fuelwood, unless repressive controls were imposed.[7]

When Rural Tree Growing Does Not make Sense

Under these conditions, rural tree growing may not be a "rational" use of scarce labour or other family assets. Some tree technologies increase the labour burden compared with agricultural cropping: notably tree alley cropping and other systems which require intensive management to yield multiple outputs. If trees demand significantly more labour than cropping (or herding), and if they do not pay dividends quickly, there may be little interest in them. Generally speaking, interest in tree growing is most likely: (1) in areas of extreme environmental degradation, where food crop yields are typically low, and (2) in high potential agricultural areas where vegetation growth rates can be very rapid and trees may yield benefits such as fuelwood and small wood products after only a year or two. However, in every location the introduction of tree growing and management has to be sensitive both to competing labour needs and local perceptions of the value of short-term versus long-term benefits.

A second type of constraint on tree projects is that short-term costs can be excessive. This is often the case with community forestry to support soil conservation, where the benefits may be large in the long term but the immediate costs can be very high. Writing about hillside closure to cattle to promote natural woody regeneration around a village in the Welo region of Ethiopia, Hultin concisely summed up this dilemma:[8] "most solutions involve short term costs which are perceived as prohibitive in relation to the stated long term benefits and – particularly in a physical environment as hazardous as that of Welo – in relation to the risk-taking involved." The costs in this case may have been difficult for outsiders to recognize and included:

- loss of pasture, especially in the rainy season(s), since grazing was restricted to valley-bottom sites formerly used mainly in the dry season. Grazing in these parasite-infested swamps reduces livestock growth
- reduced opportunities for spreading risk by cultivating a range of plots on different soil types at different altitudes. The same losses may also apply to field terracing for soil conservation and tree-planting programmes.

Food-for-work payments highlight the classic economic trade-offs which rural people make between investments in labour, etc., and short-term versus long-term benefits. If people's felt needs, whether in the short or long term, are not addressed by the intervention, they will see no reason to participate in (contribute their labour to) the programme beyond what is necessary to earn the food payments.[8] For example, in the Konso Development Programme in south Ethiopia (see Rural Case 5), where food payments were initially made for tree planting, farmers in areas where there was a shortage of suitable land abandoned the trees once they were planted and returned to crop cultivation, at least on the better sites. They clearly saw that the future value of the trees was less than the immediate value of the food payments.

The problem here is rarely one of lack of information about long-term benefits, but of the pressing needs of short-term survival. The solution may be to compensate the farmer for short-term loss by, for example, providing advice about, and planting materials for, appropriate agroforestry technologies, such as perennial fodder plants to stabilize soil conservation bunds, or tree species which are known to increase soil fertility and crop yields. Even though rigorous quantitative analyses of costs and benefits of such technologies have yet to appear, the general point remains: "conservation measures can only support, not substitute for, rural development'.[9]

When Rural Tree Growing Does Make Sense

In the previous chapters we looked at a number of advantages of trees as a part of rural livelihood systems. Here we look at some other advantages which are more closely related to household economics and the rural squeeze.

Trees as cash crops

Where there are well-developed markets for wood products, whether locally or in nearby urban areas, tree growing may well be a viable and attractive means of earning cash. Generally speaking, it is most suited to farmers who are short of capital or labour, or both, since tree management for saleable output tends to be cheaper to start and is less labour intensive than cash cropping.

In many rural areas of Africa there has been a rapid growth in markets for wood products, particularly for poles, fuelwood and charcoal. However, the causal link which is often made between the emergence of such markets and the physical scarcity of wood resources is usually spurious. The emergence of wood markets tends to be associated much more with a general commoditization of the rural economy, and with associated changes in farmer production objectives, than with physical scarcity of wood resources.[10] Indeed, once one recognizes the importance of income generation as the primary production objective for many poor farmers, the whole frame of reference of the debate about fuelwood self-sufficiency is changed.[1]

Consider, for instance, the situation in the Kakamega district of western Kenya, which is also examined in Rural Cases 9 and 14. Markets for poles and sticks for "mud-and wattle" construction are much more highly developed than for woodfuels, purchases of which are made by only a small proportion of rural people.[11] Eucalyptus and black wattle (*Acacia mearnsii*) trees are grown in farm woodlots as cash tree crops, producing construction poles and sticks, most frequently by poor farmers who are unable to meet their basic food needs and for whom tree crops are a principal source of farm income.[1]

The importance of these woodlots varies considerably from place to place. In the Vihiga area, for instance, average farm size is only about 0.6 hectares yet as much as a quarter of the area is under eucalyptus woodlots.[12] Why should farmers in one location be so interested in growing woodlots while those in other locations are not? One reason is that Vihiga and other south-eastern areas of the district are close to the markets of Kisumu and Kakamega towns. Men have learnt that if they grow eucalyptus on a five- to seven-year rotation, they can easily sell the stem wood as poles.[11]

However, this preference is also conditioned by various concerns within the household economy, including the availability of capital and labour, attitudes to risk management, and the need for cash income. Although gross returns per hectare in this area are considerably lower from tree growing than from other agricultural crops, the latter often require substantial investments which farmers cannot meet. Woodlots, on the other hand, require very small initial expenditures.[1,13] Tree crops can also be much less labour-intensive than agricultural cash crops: in western Kenya, returns to labour from pole production have been estimated to be roughly twice that of maize production.[13] This can be a crucial matter in view of the widespread male out-migration in search of wage employment.[4] Lastly, as Rural Case 14 makes clear, differing attitudes to short- versus long-term gains will affect whether woodlots are managed to produce poles every few years or smaller wood pieces such as "fuelsticks" more or less every year.

Yet these decisions may still leave unresolved – or even intensify – the problem of access to wood for use as fuel. In Kakamega the fact that wood is largely commercialized means that trees are valued for outputs of higher value than fuel for the household's own use. Gender divisions over these valuations are critical: men control access to trees and determine their uses (for sale), while women are put in the ironic position of either having to purchase fuelwood (when they can afford to) or resorting to inferior biomass fuels like crop residues, even though under male management the volume of wood resources on farms is actually increasing.

Trees for risk management

Seasonality
One potential advantage of farm trees is that they may provide a continuous flow of products throughout the year.[14] If so, they can be an important aspect of household risk management, since needs and the cash to meet them usually vary from one season or month to the next. Trees can also help to even out labour demands across the agricultural calendar, since work on tree crops can be scheduled outside peak labour periods. Alternatively, they can provide important opportunities for earning cash in the slack labour season; for example, by cropping for firewood or making charcoal for sale.

Trees as savings
Trees have also been seen as a way of meeting contingencies, such as

bridewealth or other costs incurred in marriage, funeral costs (which are often high and unexpected) or primary schooling.[15] Tree growing can generate cash with very little cash outlay, which means the benefits are relatively biased towards the resource-poor farmer. In the tropics, the "rate of interest" on tree assets tends to be very favourable because of the high potential growth rates of trees and shrubs. Although the disposal of such trees is generally a last resort, it can be an extremely important way of reducing vulnerability for the rural poor. Unlike other forms of savings, such as livestock, the cost of building up one's assets again is very low, while (if coppiced) trees go on yielding after they have been "cashed":

> ... in many rural areas of the Third World, costs of meeting contingencies have risen at the same time as traditional means of meeting them have weakened. Simultaneously, the market for trees and tree products has expanded and their value has risen. Together these trends raise and sharpen questions about the past use and future potential of trees as savings banks for the rural poor to help them meet contingencies.[15]

Several African examples bear out this point. In Benin, members of a co-operative rented land to plant trees as provision for their old age when they would be unable to do heavy fieldwork.[16] In Kenya, numerous examples are found in the literature, such as the sale of exotic species (eucalyptus, cypress and pine) in Kakamega district to pay school fees.[17] School fees were also the main reason for the sale of charcoal among the Mbere;[18] another was the purchase of food when the rains fail. In the coastal district of Kilifi palms are sold or pledged, either with or without the land on which they grow, to cover contingencies like marriage and bridewealth.[19]

This section has put the case that from the farmer's perspective – the view from below – rural tree growing may or may not make economic sense. Where it does make sense for individual farmers, however, rural tree growing, including much of what is called "agroforestry", is often also a good economic bet from the perspective of the outside donor community.

TENURE AND RIGHTS

Secure tenure over land and trees, or clear rights to their use, are obviously of crucial importance as incentives for rural people to improve these resources and hence gain benefits from them. This is particularly true for the rural poor and for activities which take a long time to produce benefits, such as many forms of tree growing.

However, rights and rules of tenure are complicated because they have at least three dimensions: people, time and space.[20] This complexity is nicely illustrated by the traditional uses of gum arabic groves in Sudan, where different people use the same space for different purposes at different times. Herders bring in animals to feed on ground cover, new seedlings and lower branches. Local farmers collect dead wood for fuel. Merchants may purchase collection rights to the gum from the tree owners. The land is usually part of a larger group's communal holdings. All of these groups seek to maintain their rights. If the land use pattern changes, for whatever reason, then a range of different interests may be affected and will need to be satisfied.

Before reviewing some of the documented examples of African land and tree tenure practices, the following general points need to be emphasized:

- tenure rules are dynamic, since they are the result of (dynamic) social relations.
- pure tenure types (freehold private tenure or strictly communal) hardly exist, especially today as state institutions are assuming tenurial roles in order to manage, develop and protect natural resources
- trees can create tenure in land and are thus a potential source of friction
- tree and land tenure frequently do not coincide; that is, people may have rights to trees or tree products on land which they do not themselves own, or to which they have no direct rights. Where there is a division of authority over land and trees, they are rarely managed as an integrated unit.

Tenure and rights change over time, most obviously when land is privatized or put in common ownership, or when a village changes its rules over access to common property resources. For example, the privatization of land in some parts of Kenya has resulted in reduced access rights to trees for many people,[21] while the long-drawn-out process of privatization has itself produced insecurities which have the same effect. Preparations for the private registration of land, or even a rumour of it, can cause the interruption or withdrawal of conventional rights. Ben Wisner and colleagues recently observed that these rights tend to disappear in the following order, in which rights over trees or their use figure prominently:[2,22]

1. building of houses
2. planting of trees
3. planting of annual crops
4. grazing of livestock
5. cutting of firewood for sale
6. cutting of firewood for domestic use
7. picking up fallen branches for domestic use
8. placing beehives in trees
9. crossing land.

Because of this dynamism, some of the specific examples quoted here may by now have changed, but the great range and diversity of practices concerning tenure and rights remain an important feature of African land management systems. Many of the examples quoted here are drawn from the excellent review by Louise Fortmann[23] and her annotated bibliography with James Riddell.[20]

Rights: Over What Trees?

Rights of access to trees and tree products can depend on various factors, including: (1) their origin (whether self-sown or planted); (2) the encompassing production system (commercial or subsistence); and (3) the uses made of them.[20]

Self-sown trees are generally community property, whereas deliberately planted trees are likely to be individually owned. In Botswana, wild fruit trees are common property unless in an individually owned field or domestic courtyard.[24] Similarly in Lesotho, wild trees are public property, controlled by the chief. In Zambia anyone can pick wild fruits, even in the gardens of others.[25]

With commercial production systems, rights of access to and the use of trees tend to be clear in the case of individual owners but much less so with publicly owned resources. For example, the gazetted National Forests of Niger, established during the colonial period from the 1930s to the 1960s, have been repressively controlled by the government

Forest Service. Access to forest products of any kind is forbidden, while strict rules govern the use of certain species regardless of their location. Yet the reserves are under-used and poorly managed by government foresters, so that they have become severely degraded.[26] This is just one instance of the much wider problem throughout sub-Saharan Africa of weak management of natural forests by the state, accompanied by strong commercial pressures on these open-access resources.

In subsistence production systems trees may be used freely by all in open-access "commons", or by more or less restricted groups in the case of controlled common property resources. In Mbere division of Kenya, for example, fuelwood could be collected from the land of another as long as it was not for resale and permission had been granted.[27]

In certain cases, a particular tree species can have a unique legal significance. This is frequently the case with hedgerow species, such as *Euphorbia tirucalli*, in Kenya. In Kakamega district of western Kenya, for example, it is irrelevant where a barbed-wire fence is located since the boundaries of a farm are always judged by the location of the obligatory *E. tirucalli* hedge. This species is used to mark boundaries as a strategy to avoid disputes with neighbours; the boundaries can be marked either by planting valuable trees – for example, for fruit or fodder – and letting neighbours take whatever produce extends into their property, or by planting neutral or "useless" trees so that there is no cause for dispute. *E. tirucalli* is just such a "useless" species. Although it may be used for emergency fuel, many superior fuelwood species exist. However, it is just this comparative uselessness which gives it its unique legal significance as a boundary marker.[28]

What Rights?

Rights in trees fall into five categories: the right to own or inherit; to plant; to dispose of; to use; and to exclude.[23]

The right to own trees is by no means automatic. At one extreme, regulations in Niger do not permit the individual ownership of trees.[29] At the other extreme, the Haiti Agroforestry Outreach Project was based on giving farmers unequivocal rights of tree ownership and complete control of usufruct (use rights).[30,31] These principles were embodied in the design of the project and are a major factor in its success. From the start it was made clear that:[30] "You will be the owners of any trees planted"; and "As far as we're concerned, you can cut the trees when you want". The Haiti Agroforestry Outreach Programme is included here (see Rural Case 18) even though it is not set in Africa because it contains three important elements: (1) the use of "umbrella" non-governmental organizations (NGOs) to disburse funds and manage the work of over fifty sub-contracted NGOs; (2) recognition that people who are destroying their environment are the solution and not the problem; and (3) respect for indigenous knowledge and techniques.

Nor is the right to plant an automatic one. It is particularly the case in Africa that planting a tree gives rights over land, so trees may be planted as a means of gaining or maintaining land rights. Consequently, women's groups may be prevented from taking up agroforestry as this might allow them title to land where this has not traditionally been permitted. There is at least one documented case in Tanzania of women being prevented from planting coffee by local leaders for this reason. Brain also quotes the case of the Uluguru Land Usage Scheme in Tanzania where local resistance to tree planting was attributed to fears that the state sought control of land.[32] Modifications to customary law have been made in parts of the country to avoid some of these problems.[33]

Rights of disposal are of four types: rights to destroy or clear; to lend; to lease, mortgage or pledge; and to give away or sell.

Rights of use over trees are often quite separate from land rights; indeed, the only case in the literature reviewed by Fortmann[23] in which land purchase automatically included trees was of the Kikuyu before 1903.[34] Rights of use also fall into four categories:

- to gather (e.g. smallwood; bark hanging off trees such as *Eucalyptus* species; or fungi, insects and birds' nests)
- to use standing trees (e.g. for hanging honey barrels, as among the Kamba, in Machakos district,[35] and the Tugen, in Baringo district of Kenya)
- to cut for construction poles or fuel, although this is generally frowned upon where there is no fuelwood scarcity
- to harvest produce such as fruit, nuts or pods. This may be separate from ownership rights: for example in Ghana, women are entitled to the kernels of palms owned by men.[23]

Rights which exclude access to trees are divided into *de facto* and *de jure* rights. Even where there are legal rights to exclude others from using trees, they may be very difficult to apply in communities based on systems of reciprocity and communal obligations. The capacities of individuals and institutions to enforce exclusionary rights is clearly of crucial importance – a point we shall return to in Chapter 7 when we look at forest management and the imposition of royalties on forest resources which are used by urban woodfuel traders.

Whose Rights?

This section considers some of the main groups who may hold rights to trees, to land, or to land and trees together. These include the state, various kinds of kin or non-kin groups, women (considered separately from men since they rarely enjoy equality of rights to land or trees), and those with temporary claim to land.

It is important to distinguish between the right *holders* and the right *users*. Rights to collect fuelwood, for example, may be held by all members of the community but are usually exercised mainly by women. There are exceptions, however. In Kagera region of Tanzania, fuelwood collection is traditionally men's responsibility and women are entitled to refuse to cook if their husbands do not provide sufficient firewood.[36]

The state

In sub-Saharan Africa, particularly in the Sahel, the state has tended to assume the role of "resource cop" through its government forest services. Access to forest products is often gained by bribing forest officers, or simply by avoiding them. The forest guards are generally too thin on the ground to enforce any prohibitive access regulations.

In Niger, for instance, the Forestry Code which dates from colonial legislation remains an enormous barrier to the sustainable management of wood resources; for example, by the Hausa and Bugaaje people in Zinder region.[37]

Under this code the state Forest Service governs access to some fifteen valuable tree species, including *Acacia albida*, regardless of location, and to all gazetted forest reserves. As a result, and even despite the relative inefficacy of the forest guard force, the remaining "rough" (non-protected) wood stock is becoming severely degraded in places. Attempts at reafforestation through a state-run village woodlot programme have failed because villagers remained convinced – with good reason – that the woodlots would not benefit them. The programme failed to address the most fundamental of villagers' concerns over trees: their rights as users of those trees. There is now growing pressure from rural people for the *de facto* privatization of wood resources: as wood resources decline, people are attempting to assert personal rights to trees growing on their own land.

An innovative and so far largely successful attempt to reverse this situation in Niger is the Forestry and Land Use Planning (FLUP) project in the Guesselbodi reserve forest. Its basic approach is to organize local people into a co-operative to manage part of the state forest for sustained yields of firewood, grazing and hay. The substantial revenues are shared by the villagers and the Forest Service. This experiment in giving local people control over and benefits from state forests is outlined in Rural Case 19 and is proving to be far more cost effective, and more effective at meeting people's needs, than conventional large-scale plantations in the Sahel.[38]

Groups

Various kinds of groups may hold tenure rights over trees or land, including kin groups or non-kin groups such as a forest co-operative – as in the Guesselbodi example cited above. The group may also have the purpose of defending villagers' rights over those of outsiders. Such rights are notoriously difficult to enforce, for instance in areas where urban charcoal cutters invade common land traditionally used by a village, or where pastoral groups pass through periodically with their animals.

Um Inderaba village in northern Sudan, for instance, lies on just such a livestock route (see Rural Case 11). The villagers were very concerned about land degradation, caused partly by the extra pressures on their resources from the passing livestock herders. An innovative community forestry project enabled the villagers to plant trees and fence off an area to allow the natural regeneration of woody vegetation. As a result, the villagers were enabled to enforce their common property rights to the planted and protected trees through a newly created commitment and vigilance.

Two interventions in Ethiopia illustrate the importance of establishing secure group rights to common property resources. The first is the Borkana Catchment pilot project in Welo region, aimed at increasing ground cover on eroded slopes, first by reafforestation and later by hillside closure. The latter method produced impressive natural revegetation in two years. However, in the absence of an acceptable group management plan for this new common property resource, local people were not enthusiastic about the initiative, regarding exclosure as a threat to their grazing rights. Popular involvement was stifled by the lack of secure rights and gains for local people.

The second Ethiopian intervention is the forestry component of the Norwegian-supported Konso Development Programme in south Ethiopia, described in Rural Case 5. In this integrated community development programme, food-for-work was initially used as an incentive to get local people to plant multipurpose trees. In some places people clearly anticipated no further benefits from the trees beyond the food payments, and saw no reason to look after the "project's" trees. Instead they grew crops on the better sites. It was not until the local peasant associations assumed full responsibility for the trees, and clear rights to their use were established, that people became interested in trees for their direct benefits and committed themselves to their care and protection. Grazing was restricted by means of fencing with thorny bushes and live bush fences which people planted.

Women

Women and their rights to land or trees have often been ignored in forestry project design, while project evaluations frequently draw attention to their low participation even in projects which are otherwise considered to be highly successful. This is precisely the case with the Majjia Valley Windbreak project in Niger (see Rural Case 12), where women have suffered the loss of grazing rights for sheep and goats.

Other projects have specifically set out not only to safeguard local women's existing rights, but to build on opportunities to strengthen them. The KWDP achieved this very well through its *Moto Mwaka* ("firewood all the year round") mass awareness campaign (see Rural Case 9). While women had access to land, they did not have access to the trees growing on it, and were finding it increasingly difficult to fulfil their responsibility for collecting fuelwood from common land. In its extension programme the KWDP managed to encourage the planting of trees by women – traditionally the exclusive preserve of men – by using a shrubby species (*Sesbania sesban*) which is not categorized as a tree with respect to planting or use rights and can be intercropped very successfully with food crops.[17] This approach has considerable potential for safeguarding fuelwood supply through strengthening the rights of women at household level. This is particularly important in some of the poorest parts of Kenya, for example, where as many as 30 per cent of all rural households are headed by women.[2]

The extent and nature of women's rights vary considerably throughout southern and East Africa, although it is frequently the case that women have no rights to land.[23] Women in Lesotho[39] and Botswana are legal minors although they may own land. Hammer writes that it is rare to find women owning land in the Sudan[40] and that female heirs get half the amount that male heirs get.[41] On the other hand, Tanner found nearly as many women as men owned palm trees on the Tanzanian coast, reflecting the influence of Islamic law.[42]

In Burkina Faso, women's lack of land rights may explain why, historically, few women have planted trees. The government of Burkina Faso declared in 1984 that although all land belongs to the government, rights of use remain with Burkinabé citizens. It remains to be seen whether this policy change will improve women's tenure rights, or affect people's willingness to plant trees, although the revolutionary government has stressed the importance of mobilizing women to take part in all development activities, including tree planting.[43,44]

Temporary claimants to land or trees

This category include groups such as tenants, lessors and pledgees, squatters and borrowers. Often the person who plants a tree is its owner, even where s/he does not own the land on which it stands. On the other hand, rights to trees on land to which one has only temporary access may be restricted owing to the possibility of using trees to establish a more permanent claim to the land. For example, Kikuyu tenants in the last century could not fell trees without the landlord's permission.[34] Similarly, Swanson reports that among the Gourmantche of Burkina Faso, land borrowers are not permitted to plant trees since this would establish inheritable rights in the trees and, in practice, in the land shaded by them. Out of a sample of 311 people, 91 per cent said they would not permit the planting of mango on their land by borrowers.[45]

An extreme case of people with only temporary claim to land or trees is that of *refugees*, who have been dispossessed of their land and homes usually because of war or severe drought. In Africa, where refugees are now thought to number well over 10 million, the two conditions frequently coincide, with war a major cause of famine in the face of drought. Despite living in fixed camps, often for many years, refugees tend to have no security of residence or land tenure whatsoever.

This lack of security poses particular problems for community development projects in refugee camps which try to encourage people to plant trees for fuelwood provision and environmental rehabilitation. Although several such projects have been attempted, one which achieved unusual success was the InterChurch Response "Step" Plan for refugee forestry in Somalia (see Rural Case 20).

Its innovative approach was to involve people in tree planting by means of a simple, step-by-step training course and small payments for work by the participants. Cash payments for seedlings raised, and simple tools as payment for planting and land preparation, provided immediate benefits to the refugees, who lacked any incentive to wait for longer-term benefits as they did not expect to be around when the trees matured. The project was by no means "participatory", in the true sense of the word, nor were its forestry techniques particularly innovative for a dryland environment. However, in the extremely harsh conditions under which it was operating, and given the particular problems of refugee insecurity, the project probably achieved about as much as it could with its innovative idea of providing frequent rewards to offset the lack of interest in long-term gains.

Enforcement of Rights

The most costly and least effective mechanism for enforcing rights, on the whole, is to rely on state power. The difficulty of enforcing prohibitions on the use of state forests in Niger and elsewhere has already been noted, while the problems encountered with the forced destocking of cattle in the HADO soil conservation project in Tanzania are discussed in Chapter 1 and Rural Case 13. In the latter case, the rights which were denied were so basic to livelihoods that they were both ineffective and imposed considerable social costs.

The difficulty of enforcing such measures is also illustrated by the example of Ethiopia, where following the 1975 revolution all land became the collective property of the Ethiopian people. The former Forest and Wildlife Conservation and Development Authority was made responsible for the control of forest areas larger than 80 hectares, but because it lacked the necessary legal authority to enforce its control, the forests effectively remained open-access resources to be exploited as individual communities thought fit.[9] At the same time, the newly formed peasant associations were charged with conserving their local forests and exhorted to plant trees in designated forest reserves. But because the scheme lacked both the real backing of local people and the institutional means to make enforcement effective, land degradation is continuing.

Community sanctions are the ideal mechanism for the "enforcement" – through voluntary agreement – of rights within fairly small and defined communities. They are more likely to put people's priorities first, to be effective, and to be sustainable in economic and social terms. Table 3.1 summarizes the many types of possible control systems which are used in traditional forest management by local communities.

Table 3.1: Control Systems Used in Traditional Forest Management

basis of group rules	examples
1. Harvesting only selected components	• Trees: timber, fuelwood, fruit, nuts, seeds, honey, leaf fodder, fibre, leaf mulch, other minor products (gums, resins, dyes, liquor, plate leaves, etc.) • Grass: fodder, thatching, rope • Other wild plants: medicinal herbs, food, bamboos, etc. • Other cultivated plants: maize, millet, wheat, potatoes, other vegetables, fruit, etc. • Wildlife: animals, birds, bees, other insects, etc.
2. Harvesting according to condition of product	• Stage of growth, maturity, alive or dead • Size, shape • Plant density, spacing • Season (flowering, leaves fallen, etc.) • Part: branch, stem, shoot, flower
3. Limiting amount of product	• By time: season, days, year, several years • By quantity: of trees, headloads, baskets, animals • By tool: sickles, saws, axes • By area: zoning, blocks, types of terrain, altitude • By payment: task, kind, food, liquor, manure • By agency: women, children, hired labour, contractor, type of animal
4. Using social means for protecting area	• By watcher: paid in grains or cash • By rotational guard duty • By voluntary group action • By making mandatory the use of herders to watch animals

Source: Arnold and Campbell (1985).[46]

The critical issue is not so much what rules are applied but the strength of community institutions which set the rules and ensure that they remain effective. Community sanctions are most likely to arise spontaneously and work where a cohesive social and administrative structure exists. This will tend to be in relatively isolated places which are little affected by migration either to or from the area, as with the examples of voluntary exclosure of cattle from grazing areas to allow natural woody regeneration in Shinyanga, Tanzania, and the initiatives taken in remote villages in northern Sudan. (These are described in Chapter 1 and Rural Cases 1 and 11.)

At the other extreme, totally transient communities such as refugee settlements are most unlikely to develop such rules themselves. Between these extremes lies the great majority of communities with a variable social mix of insiders and outsiders, or those where community stability has not been achieved or has been upset. Here, social sanctions surrounding access to natural resources are unlikely to arise spontaneously but there may well be a willingness to develop them. The extent to which outsiders can help do this is, as yet, largely unknown. It may be possible to support and strengthen existing structures or institutions, or create new ones where the basis for rule making is totally lacking. The real problems arise where there are vestiges of institutions that can neither be built upon nor ignored.

GENDER ROLES

Divisions between men and women in access to natural resources, and in their management and use, are common in African land management systems. In the previous section, "Tenure and rights", we discussed briefly one aspect of these divisions: the issue of women's rights to land and trees – and hence to woodfuel – and how they differ from those of men. Another example that we might have chosen is the difference in the reasons for the widespread practice of burning bushland – a frequent target of public education campaigns in Africa because, to the outsider, it appears to be so destructive. Whereas men may burn fields to clear them or burn pastures to promote the growth of grass for livestock, women often burn to stimulate the fruit and nut production of trees.[44]

But gender divisions are much wider and run much deeper than these examples suggest, while it is obvious that unless they are addressed forcefully yet sensitively, development efforts will usually be ineffective.

The international development community has, of course, begun to recognize this point with much talk and some action about increasing "women's participation" in development. All too often, though, participation is limited to small aspects of a project, such as ensuring that more women are used for labour so that they earn cash of their own. This falls far short of true participation by women, which at the very least must include their active involvement in community discussions which identify problems and help to define project objectives and how these should be met, as well as in management during the project cycle. Efforts must also be made to ensure not only that women share equally in the benefits of development but that development activities are themselves designed to achieve this objective.

Consider, for example, the many projects designed to ease the time burdens for women of collecting firewood. This may well be a serious problem, but there is a danger of seeing it purely in terms of labour issues within households. Women are not simply firewood consumers and collectors. They play a specific and integral part both in household consumption and survival strategies, and in the management and utilization of trees and shrubs within overall land management strategies. A comprehensive view of these strategies in all their complexity and diversity demands a clear understanding of the interactions between women and forests and trees in farming systems.[44,47]

Forestry development initiatives must therefore not just "consider" women but aim at giving them equality with men in control over resources, in decision making over resource production and use, and in empowerment to evolve self-directed problem-solving strategies. For this reason, projects aimed specifically *at* women have often been unsuccessful, as they tend to ignore the broader social reasons why women are second to men in these concerns. More promising are efforts that emphasize involvement of all family members, with an explicit focus on reaching women.

Worldwide, the main obstacle to equal participation by women in forestry development has been the narrow focus on "trees" rather than "people and trees". If a people-centred approach to forestry and wider aspects of rural land management can be developed, almost by definition women as well as men will gain from it.

Chapter 4

Meeting the Constraints

So far we have argued the strong case for voluntary local initiative and involvement in development efforts, not least for rural forestry and fuelwood. The case can be summarized as follows. First, the most successful interventions are usually those with visible, widespread, local involvement in their design and implementation. Projects with weak local involvement have often failed or been too expensive. Second, government or donor resources alone cannot meet all the tree-related needs of the rural poor: people must do as much as possible for themselves. Third, the diversity and complexity of rural people's livelihoods and rural landscapes in Africa make it essential that interventions are site-specific. Altogether, this implies a need for outsiders to learn from and work with local communities.

In this chapter we ask the question: how? What institutions, or institutional partnerships, are needed to effect more participatory approaches? How can government or donor "projects" increase their participation with local people? And what does that fashionable word "participation" itself *mean*? To enter this jungle of questions, we start on the outside – at the "top", with central governments – and work our way through to more decentralized and people-centred approaches, investigating how these can be matched with the "projects" of outsiders.

GOVERNMENTS AND THE LEGACY OF HISTORY

Many African states have inherited government structures and attitudes from colonial times which are wholly inappropriate for the tasks they need to perform. Forestry departments in particular are often geared towards conserving forest resources rather than their sustainable management and use to meet people's various needs. Indeed, in several countries foresters even continue to play the role of "resource cops" which they took over from colonial rule, often nowadays with the support of major international donor agencies who give higher priority to restrictive protection than the mobilization of people's efforts to secure a more sustainable development.

In Chapter 1 we saw how such attitudes have dogged efforts in Tanzania and Niger to get local people to plant trees – but these were merely selected cases out of many in a large number of countries. In Tanzania, for instance (see Rural Cases 13 and 17), the creation of forest reserves has entailed the evacuation of the local people and the policed denial of access to what were formerly common property or open-access resources; soil erosion programmes have enforced the destocking of cattle and hillside closures; even the village woodlot programme with its best intentions of decentralized decision making was in practice controlled from the centre and driven by political needs at the highest levels to meet popular targets. For the local forester, each new village woodlot, whether or not it met people's real needs, became a counter in the game of securing promotion or the approval of superiors.

This kind of situation, we repeat, is not unique to that country – or to governments. Setting physical targets as the goal of development programmes – rather than aiming to set in train less tangible processes of human development – is a common feature of most hierarchical, bureaucratic and technically "rationalist" institutions; that is, of most government and many large donor organizations. It helps to explain, for instance, why there are such strong biases in agricultural and forestry programmes to work with larger or more "progressive" farmers, since they are more likely than resource-poor farmers to deliver the goods to meet preconceived targets.

Writing in the context of protection forestry, Blair and Olpadwala have summarized the change that is needed to correct such institutional biases:

> the problem of transforming forestry services from regulatory agencies to [rural development] extension organisations ... is often depicted in terms of turning forest guards into extension agents, or changing policemen into salesmen, and is rightly considered a difficult goal to achieve, albeit one well worth striving for.[1]

Added to this institutional legacy is the lack of co-ordination which is often found between specialist government institutions which ought to be working together towards the same objectives. This is such a familiar complaint, in the North and South alike, that we need not dwell on it.

However, there are several examples from Africa where two or more government agencies have major interests in forestry but adopt different and conflicting objectives: one arm, for instance, emphasizing trees for soil conservation, the other, trees for multiple outputs. In some countries one finds that villagers are visited by three or more different types of extension worker – for agriculture, livestock, forestry, and water, for instance – each providing advice which is at best independent of the others and, at worst, in complete conflict.

A third historical legacy stems from another form of uncoordinated specialism: what Robert Chambers has called "normal [development] professionalism".[2] This runs through African government bureaucracies just as it does through the governments and development agencies of the North. It refers to the thinking, values, methods and behaviour dominant in and rewarded by professions and disciplines, and reflecting "core" or "first" biases. These conservative biases are towards the needs and interests of the better-off and the urban; towards high technology, quantification and men. Within the rural development community, disciplines such as agronomy, forestry, community development, administration, health, water development and animal husbandry divide up rural livelihood systems into "sectors". Specialization is one of the defences of normal professionalism, where for example "foresters stick to trees, and moreover to trees in the forests and forest plantations which they control … [and agronomists] stick to crops, those in which they have specialized".[2]

A major corollary of normal professionalism is the devaluation of popular knowledge, as we have discussed in Chapter 1. Many professionals cannot believe that poor people can know anything of consequence. To counter such "first" biases Chambers makes a powerful case for a reversal of the "first" paradigm, and for "last thinking" to be put first. The "last-first" paradigm includes learning from the poor and putting greater emphasis on decentralization, empowerment, local initiative and diversity. Development must become a learning process, not a march across a blueprint, since successful development is flexible, iterative and adaptive. "New professionalism questions conventional methodologies for appraisal, research, training and managing bureaucracies, and exploits gaps between disciplines as opportunities for the poor. To achieve reversals on a massive scale is now perhaps the greatest challenge facing the development professions."[2]

The legacy of the past – anachronistic institutions, ill-advised and poorly specified development efforts, and the professionalization of development itself – has a major bearing on what can and what cannot be achieved today. Where outsiders become involved in enhancing local people's capacities to strengthen and enrich land management strategies, their efforts inevitably depend upon the acceptance and trust of local people. But if local people's prior experience of outsiders has been a catalogue of virtually unmitigated disasters or "top–down" approaches, it will take a long time, a lot of patience, and radically different approaches on the part of outsiders to achieve anything at all.

In order to tackle woodfuel problems and broader resource and livelihood issues in a holistic and relevant way, there is a need for new institutional mechanisms to facilitate indigenous initiatives at the local level; joint policies which address common causes of problems; new alliances between separate or competing agencies; and new forms of intervention that match up to people's needs and demands.

In this chapter we consider a few possible mechanisms for such "institutional innovations", again using case histories as illustrations wherever possible.

CROSSING INSTITUTIONAL BRIDGES

Just as rural people's livelihood strategies mimic the complexity and diversity of African landscapes, so successful initiatives in wood resource management must take a correspondingly eclectic approach. They must respond to the multiple uses people have for trees and shrubs, and address whatever needs and priorities are predominant in each place. This may involve either a more integrated approach between different agencies or objectives, or working "through the back door" by meeting a primary objective indirectly.

One way of taking an indirect approach is to incorporate trees into projects whose principal objectives lie elsewhere. This idea has been examined in Kenya for several projects including those aimed at dairy development, rural road building, and integrated rural development (see Rural Case 21). Similar potentials were identified in poultry development programmes and small-scale irrigation schemes.

In the Gutu district of Zimbabwe, an integrated approach to wood resource management is being attempted through the Co-Ordinated Agricultural and Rural Development (CARD) Programme.[3] This is the first major programme in the nine southern African countries of the

SADCC region to incorporate a woody plant focus into an integrated rural development project. A major objective of the programme is to improve both arable and pasture management by growing woody plants, such as trees for fodder production. It is too early yet to say whether or not this will be successful, but at least the principle of taking an integrated approach is central to the design of the programme.

In Sudan, the inclusion of forestry in irrigated agricultural schemes dates back to the 1940s but has had a chequered history.[4] Agricultural planners have often been hostile to foresters, despite the manifest advantages of having more trees on the irrigated lands. Broad estimates suggest that by planting trees on canal banks, marginal lands, and only 5 per cent of the agricultural land in these schemes (in the form of shelterbelts and border plantings), over 30 per cent of urban fuelwood demand and over 10 per cent of total fuelwood demand in central Sudan could be provided for on a sustained yield basis.[4]

However, since 1983 a new irrigated agricultural scheme at Seleit, only 10 kilometres north of Khartoum, seems to be resolving this tension simply by the device of making foresters answerable to the scheme management rather than the local forestry authority. In this way, agriculturalists and foresters co-operate and integrate their activities to meet common objectives. For example, the intercropping of eucalyptus trees with *karkedeh*, a hibiscus species used for drinks in Egypt and Sudan, raised substantial revenue for the scheme through the sale of the first *karkedeh* crop.[4]

In each of these examples the principle remains the same: wherever trees or shrubs are or can be used, identify the whole range of local uses of and needs for wood resources, establish which species and forms of management are most appropriate and economic (in combination with other demands on land or labour), and then ensure that the institutional arrangements are such that the most favourable options for local people are pursued.

INSTITUTIONAL PARTNERSHIPS

In some countries appropriate local institutions already exist and may only need strengthening to expand the scope of their activities – although this is often easier said than done. In other countries the right institutions may need to be created, or new partnerships formed. In this section we examine various kinds of institutional partnership that have the potential to "reach the people".

Perhaps the greatest challenge in creating such partnerships is to achieve the best mix between local "grassroots" development organizations and government. The best of the former are generally strong on social commitment, motivation and the ability to innovate. They are the essential end-links in the design of technologies or their delivery to local target groups to meet specific constraints and needs and to implement truly appropriate change by giving local people the power to control and benefit from these technologies.

At the other end of the pyramid is central government. It has strengths too. It can have power and reach on the larger issues: for example, fostering an innovative climate; funding research, development and demonstration; supporting integrated development projects and extension services; and, not least, legislating on land ownership and distribution, rights to resource use, controls on resource abuse, or economic issues such as food pricing.

The problem is how best to combine the power of the big with the sensitivity of the small in order to foster and deliver rapidly innovations that people need and will benefit from equitably and sustainably. "The search for and growth of effective partnerships of this kind are surely among the most urgent challenges of development."[5]

Of the institutional partnerships which we examine, some are between government or other "official" structures, others are between NGOs, and yet others involve links across government and voluntary agencies. Clearly, the precise institutional structure will need to be tailored to suit specific circumstances and the strengths (or weaknesses) of existing institutions. The various institutions which may be involved in such partnerships are categorized in a highly summarized fashion in Table 4.1.

BUILDING ON LOCAL ORGANIZATIONS

When the American private voluntary organization CARE first became involved with agroforestry projects in Kenya some six years ago, they found it difficult to identify appropriate local structures to use as extension agencies and felt it necessary to create new ones. Their basic objective was to act as a local catalyst and to hand over full responsibility for the project to a sustainable local structure, as well as

Table 4.1: Framework for Institutional Partnerships

type of institution	scale of operation		
	community/local level	regional/national level	international level
Government or other "official" body	• Local authorities/ village councils • Political party structure • Locally-based extension services • Farmer co-operatives • "Projects"	• Government line ministries • Parastatal bodies	• Multilateral and bilateral agencies • UN structure
Non-governmental (NGOs)	• Local organizations: e.g. self-help groups, women's groups, church-based groups • Work groups • User group organizations • "Projects"	• Co-ordinating NGOs e.g. KENGO, Six "S" Association and various consortia of relief agencies • National/regional NGOs	• Private voluntary organizations: e.g. OXFAM, CARE, World Neighbors

stimulating farmer-to-farmer extension. However, such a hand-over had to be carefully planned from the outset.

Table 4.2 identifies some institutions which achieve effective out-reach at the local level in their respective countries and which could form a model for local organizations dealing with natural resource management elsewhere. There may be an important role here for outsiders – such as donor organizations – to carry out "rapid institutional appraisals" to identify opportunities for creating such institutions or "piggy-backing" on existing ones. In Uganda, for instance, the Danish aid agency DANIDA funds studies in fifteen districts to discover just these sorts of local mechanisms for innovative new forms of extension.[6]

Work Groups

Probably the simplest and most widespread of all traditional local institutions is the work group, known in West Africa as the *dokpwe* and described by Roland Bunch as follows:[7]

> A group of farmers works together, one day on one of the farmers' fields and the next day on another's ... Through such a simple

Table 4.2: Possible Institutions as Vehicles for Facilitating Local Initiatives in Tree and Shrub Management

country	institution	comments/activities
Burkina Faso	*Naam* groups	Based on traditional Mossi self-help groups. Six "S" Association as an umbrella organization to co-ordinate and provide back-up (Rural Case 22)
Tanzania	Primary health care The Party	Excellent outreach Highly organized at local level, even achieving inter-sectorial co-ordination of activities in some cases
Ethiopia	Peasant Associations Service co-operatives	OXFAM trying an innovative approach to participatory planning
Sudan	Village councils	Successful at Um Inderaba village in the northern Kordofan (Rural Case 11)
Uganda	National Resistance Committees	Play a similar role in Uganda to the Party in Tanzania
Kenya	Various NGOs	Very strong NGO community relative to other countries
Mali	Development Committees	Achieve good outreach, have knowledge of local conditions and can reach decisions quickly. Already co-ordinate government technical services at local level

institution, villagers can learn to work together, get to know each other, share tools, do slow-moving large-scale jobs without becoming discouraged by the apparent lack of progress, and learn from each other about agricultural innovations. Starting with a simple institution like the work group, people can learn to work together in progressively more complicated institutions, such as community action groups. ... The possibilities are as numerous as the needs.

User Groups

Common property resources such as grazing lands or woodland are
exploited by definable user groups. A "user group organization" has
been defined as a minimal form of organization that may be able to
undertake common property resource management.[8] The sanctioning
authority for the group will frequently be the village council. Its
activities may include: making and enforcing rules of entry and exit to
the resource by organization members; regulating patterns of use of the
resource; assessing and imposing penalties on outsiders and insiders for
rule infringements; developing means for resolving conflict; and
organizing the collective improvement of the common property
resource.[8]

The ability to organize communally in this way will be severely
limited unless the community has lived near the common resource for
a relatively long time, and under conditions of moderate scarcity.
However, where these conditions apply it is highly likely that some type
of user group organization will have already emerged in the community.[9] In the Shinyanga region of Tanzania, for example, there is clear
evidence that user groups can effectively regulate the use of common
property resources, notably the traditional *Ngitiri* grazing reserves (see
Rural Case 1).

Usually it is far better if outside agencies support such established
groups, if necessary, rather than try to establish new institutions in the
same place. All too often, organizations are created at village level (and
at higher levels) that are little more than skeletons. The mere existence
of the framework of an institution does not "bring it to life" in terms of
sharing a sense of common purpose on the part of the people involved.
"The need for an institution should not arise from the development
agency, but from among the people .... If there is no definite felt need,
no institution should be formed."[7]

Co-Operatives

Co-operatives are popular with outside agencies as they seem such an
obvious way of encouraging villagers to work together in managing a
common resource and distributing the benefits. However, they may not
be appropriate in many cases. Co-operatives are a European institution
designed to meet European needs. They are complicated, difficult to
understand and even more difficult to run,[7] particularly if much

accounting of production and payments by individual members is required. In fact, co-operatives are usually only successful "in areas already quite far advanced in commercial development and skills".[10]

The forestry co-operative set up as part of USAID's Forestry and Land Use Planning Project in the Guesselbodi reserve forest, Niger, has been one successful attempt to create an institution of this sort (see Rural Case 19). But in this case there are particularly favourable conditions for a commercial co-operative to flourish: most importantly, the market opportunities offered by the capital city of Niamey, which is only 30 kilometres away on a good metalled road, where fodder and firewood produced by the forest fetch high prices.

Local Non-Governmental Organizations

These include community-based self-help and church-based groups and often have a strong record in securing the commitment of local people and of group members. One of their major potential advantages is their decentralized way of working, which allows them to address local diversity and incorporate local knowledge, as they tend to be familiar with local realities.

Their strength and numbers vary considerably from country to country, according to historical, political, economic and social factors. In Tanzania, for example, the NGO community is extremely limited and met with governmental disapproval until recently. By contrast, Kenya has a very wide range of locally based NGOs as well as national and international NGOs or private voluntary organizations, even among those dealing with woodfuel and forestry issues. Indeed, the latter appear to be so numerous that it might be thought they could easily come into conflict from time to time. However, KENGO (Kenyan Energy NGOs), which acts in a co-ordinating role, explains that to avoid conflict, different NGOs (1) are responsible for different geographical areas of the country, and (2) take different and complementary approaches. For instance, the Green Belt Movement distributes seedlings free, while KENGO itself is involved in training.[11]

Co-Ordinating Local Institutions

There are several reasons why the co-ordination of local organizations is a good idea. It can avoid duplication of efforts, help to pool and share skills and experience, and strengthen the spirit of small, possibly remote groups by making them part of a larger structure with greater presence and political voice.

Co-ordinating structures can also be vital as a means of channelling funds to a plethora of groups which are too small and numerous for governments or donor agencies to deal with directly. A successful example of this approach is the Haiti Agroforestry Outreach Programme (see Rural Case 18), in which USAID funds were distributed through three private voluntary organizations to over 170 locally based NGOs working throughout the country.

A second reason for co-ordinating local organizations is to foster a wider and more effective social movement. In some cases, the dynamism of a particular individual can be multiplied in this way, as in the Koumpentoum Entente in Senegal, an association of self-help village development organizations that received help and encouragement from a famous Senegalese extension worker, Mahamadou Cissoko (Rural Case 4).

In Burkina Faso, the successful *Naam* movement of village self-help development groups, founded by Bernard Lédéa Ouedrago, is co-ordinated at the national level by the Six "S" Association (see Rural Case 22). This umbrella organization not only provides support to local groups by linking them together but also acts as a channel for international funds to each village group. A key feature of this popular movement is its historical roots in the strong Mossi traditions of self-help.

NEW GOVERNMENT STRUCTURES

Government policies and structures can either hinder or facilitate local initiatives in natural resource management. Where policies stifle indigenous initiatives the most pressing need is obviously to create a more favourable policy environment. One interesting example where this has been done is the new programme known as Project Campfire, initiated by the government of Zimbabwe with the aim of involving local communities in deciding for themselves how best to manage their local resources (see Rural Case 23). Unfortunately, the programme – and the resource management co-operatives which it is helping to set up – is too recent to evaluate.

Similar kinds of policy initiative are being taken in other countries, such as Mali and Burkina Faso, where there seems to have been a major shift in both the rhetoric and the content of government approaches to the management of resources by local communities. In both countries, villagers are made *maîtres d'oeuvre* (in charge) of development activities, through a process involving consultation with village committees to identify local priorities, and the preparation of local plans, which make clear the contributions to be made by villagers, local and national government, and donor agencies.

Although at an early stage, these new developments are showing encouraging signs. In Burkina Faso in particular, there is great demand by villagers for help in establishing soil and water conservation schemes, following the successes achieved by farmers in the Yatenga region (see Rural Case 10). In Mali, the newly established National Environmental Council receives the backing of the highest political authority. It is ensured the necessary support to force changes in the policies of individual government ministries, where these are necessary to support the environmental programme.[12]

Where institutional structures thwart the incorporation of sustainable environmental management principles into general policy making, it is often because of "sector thinking". The absence of sideways linkages between different sectors (agriculture, forestry, livestock development, water development, etc.) frustrates their co-ordination. Within each sector the command chain tends to be vertical, and usually from top to bottom: decisions are made in central ministries and flow down to the local level where they are – or are not – implemented.

In Ethiopia for example, there are now two governmental departments charged with a forestry brief. One is the State Forest Conservation and Development Department, which has responsibility for the designated state forests. The other is the department of Community Forestry and Soil Conservation (CFSC). Neither agency has a clear responsibility for soil conservation, or for forestry, or for the way in which both relate to wider policy objectives, for example agricultural development.[13] The CFSC, for instance, tends to emphasize soil conservation rather than the wider set of benefits obtainable from multipurpose tree species. The result is an overall tendency towards conservation, which many farmers reject, rather than the sustainable utilization and management of wood resources.

In Kenya, efforts are being made to overcome the vertical segmentation of rural "problems" and interventions designed to alleviate them. A new Biomass Division has been created within the Ministry of Energy and Regional Development (MOERD), which has agroforestry as one of its areas of responsibility. Six Agroforestry/Energy Centres, which were set up with support from the US-based consultancy Energy/Development International under the Kenya Renewable Energy Development Project, now fall under the auspices of MOERD's Biomass Division. Each centre has responsibility for a particular agroecological zone and has the potential for addressing a wide range of biomass resource issues in a holistic way.

Decentralized Government

The desire to bring about more effective local development has led to attempts in many countries in the South – and the North – to decentralize their government structures. The expressed motivation is usually to bring government closer to the people and to foster a more favourable environment for local initiatives. But however laudable the intentions, decentralization is not without considerable practical difficulties.

In particular, effective decentralization has to involve meaningful local representation. Decentralized government institutions are not intrinsically flexible and responsive to local needs, and while they should reduce delays in implementing development activities and providing local services, as well as reducing implementation costs, the mechanisms for achieving these gains have to be thought out thoroughly beforehand. The critical issues are whether field personnel are competent, whether they have the authority to make financial decisions, and precisely how local people will be involved in development programmes.

Furthermore, much depends on the precise relationship between local and central government. Local authorities in Africa, particularly in rural areas, usually lack the funds to finance development on their own, and are often short of skilled staff in particular areas. This problem will not be solved by making local government more autonomous unless, at the same time, an effective partnership structure is created between local and central government.

One notable attempt to do this is the District Focus Strategy formulated in 1983 in Kenya.[14] Until then, all development activities were formulated in Nairobi and passed down to district level as blueprints for action. These were usually sectorial plans by individual ministries and lacked co-ordination at district level. Under the new strategy, which aims to make districts the focal points for rural development, this was all meant to change. Strong district development committees (DDCs) were created and made responsible for identifying in detail the development programme in their areas and co-ordinating the development activities of local authorities.

However, the ground had not been adequately prepared. Some observers regard the new district governments as just another form of central government and believe that authority has not been devolved far enough.[15] The ability of local authorities to contribute to local development has actually been hampered by their own internal constraints and by central government decisions. Not only are they unable to generate adequate revenue, but they are plagued by a lack of skilled and competent personnel, and by having to share with central government responsibilities which overlap or even come into conflict. In short, the "partnership" remains a top–down one, with central government holding the reins through its regulatory, rather than facilitating, authority over local councils.[15]

Government Extension Services

The extension services of line agencies dealing with agriculture, livestock, forestry, health, etc. are normally the main ways in which governments reach out to rural life. Who, then, should look after rural forestry and woodfuels? Normally these topics are the jealously guarded preserves of forestry extension services, but for two reasons agricultural extension has a much greater potential for effective outreach. First, agricultural extension workers typically outnumber their forestry counterparts in Africa by at least ten to one. Second, agricultural (and livestock) extension services deal with more fundamental issues of land management and concerns of farmers and herders than do forestry *specialists*.

The problem, then, is how to teach agricultural and/or livestock extension workers about trees as an integral part of land use production systems: an easier task, perhaps, than training foresters in agriculture and livestock management.

Non-governmental organizations can contribute here, to supplement the hard-pressed extension training colleges and curricula. The Kenyan NGO consortium KENGO, for instance, is starting to train government agricultural extension workers in agroforestry techniques, on which it has a wide experience.[11] The decision to do this rather than focus on forestry workers (who also frequently need training in agroforestry) was made because in several areas of Kenya there are simply too few forestry extension officers to achieve any real outreach; Kajiado district, for instance, has none at all.

Another form of partnership is to link apparently disparate government extension services. Community health care, in particular, often has an impressive outreach not only to villages but to the household level and, especially, to women. Tanzania is a notable example. The links between better nutrition and trees (for fruit and nuts, etc.) and between exhaustion or malnutrition and acute fuelwood scarcity, suggest that village health clinics could have a significant role in woodfuel and forestry issues. On the one hand, health workers could be trained to be alert to any signs of a locally worsening woodfuel situation, using proxy indicators as a form of "rapid appraisal" problem identification. On the other hand, they have an interest in helping villagers to plant trees to increase food production. The Saradidi Health Care Project in Siaya district of Kenya aims to integrate agroforestry extension in this way, in tandem with CARE's Agroforestry Project in the district.[6]

Finally, but not least, is the urgent need for women to become partners in extension work. The dominant role played by women in supplying fuelwood, and the gender divisions which so often disadvantage them, make it imperative that they should be at least equal to men as targets for forestry advice and interventions. Yet achieving contact with women is often a bottleneck in male-dominated extension work. In African forestry few, if any, extension workers are women. Until recently in Burkina Faso, for example, all three professional women foresters were engaged on the improved cookstove programme. But this situation is now changing, with greater use of women as forestry and rural extension agents. Similar efforts are under way elsewhere. In Senegal, for instance, female extension agents are now working with male forestry workers[16] and in the Sudan, women extension agents are receiving on-the-job training in forestry.[17]

THE EXTENSION CHALLENGE

The success of rural extension services hangs or falls on having enough appropriately skilled personnel. That seems obvious enough. Equally obvious, the supply of such people in poor countries will always be limited. The question then is how to develop enough people with the appropriate skills; this includes the question of what one means by "appropriate".

On Participation

A consensus is emerging that appropriate extension work or other types of intervention must involve "participation". Indeed, the examples of promising or "successful" local-level initiatives we refer to in this book all indicate that good participation is a key ingredient in any successful development strategy. However, if participation is to be more than just another vogue concept, we need to consider the nuts and bolts of what it really means, how to achieve it, and how to improve its quality.

Roland Bunch of the American private voluntary organization World Neighbors, in his guide to people-centred agricultural improvement *Two Ears of Corn*,[7] makes many penetrating observations about the nature of participation, based on the practical field experiences of World Neighbors over the years. Here we briefly summarize that critique.

Participation can be destructive as well as constructive. Sometimes particular individuals emerge as leaders and take control. A lack of experience in handling money and in decision making as a group often causes disagreements. The dishonest or inappropriate actions of some leaders may go unchecked because many cultures have no acceptable method of dealing with such behaviour. A failure to provide rapid, tangible benefits for the local community, even if the right things are being done, often leads people to lose interest, and whatever enthusiasm they had wears off. Such experiences may lead outsiders to believe that villagers are untrustworthy and incapable of solving problems. These kinds of participation "are destructive rather than constructive. They do not *produce* development; they *preclude* it".[7]

Constructive participation by people is the opposite of doing things for people. Through it people learn to plan, to find solutions to their problems, and to organize themselves for collective action. They can gain self-confidence, pride and the satisfaction of having made significant achievements and of being able to continue to improve their lives creatively. This is what constructive participation really means: it is the basic building block of a development approach which is about local problem solving. In turn, development itself is "a process whereby people learn to take charge of their own lives and solve their own problems".[7]

Bunch has identified a number of features which are important in increasing constructive participation within people-centred development programmes. The most important of these are to:

- create enthusiasm
- start small and simple
- be careful with the role of outsiders
- plan for the phase-out of outsiders and of the intervention itself
- teach farmers how to conduct small-scale trials
- build a leadership pyramid
- don't flaunt the moneybags
- don't try to meet all of people's needs
- remain constantly aware of the level of villager participation.

In addition, careful attention must be paid to improving the quality of participation. Constructive participation is learned gradually, and conscious and constant efforts need to be made to help people learn how to participate effectively. Early, tangible benefits for local communities are crucial to their continuing participation.

Training for Extension

These comments on participation are highly relevant to the whole issue of how to improve extension services and how extension agents can be trained and developed. There is a growing consensus that good extension depends on certain guiding principles, including:

- the need for a two-way flow of information, making the most of indigenous as well as outsider knowledge; that is, a "farmer-first-and-last" model
- the need to act as facilitator or catalyst, which implies that new forms of communication may need to be emphasized, especially *listening*
- the need to identify target groups, recognizing the importance of structural factors and constraints, such as gender roles within the household, land and tree tenure, market opportunities, etc.
- the need to develop a process- rather than an outcome-orientated approach
- the need for sufficient humility or patience not to take the locus of control, action and decision making away from the target group and individual farmer.[18] In other words; the people should "own" or control the process, not the extension staff.

Generally speaking, local people are more likely to meet these criteria than outsiders since they are more familiar with the practices, cultural traditions, constraints and opportunities of the target community and so should find it easier to work alongside them and gain their trust. However, where outsiders must be used, their work can be facilitated by several methods which help to ensure that extension work is appropriate and remains "people-centred". They include:

1. mass media, such as radio and television, as in Tanzania's "Forests Are Wealth" campaign (see Rural Case 17); an advantage of mass-media promotion is that the farmers can more easily interpret and adapt the message themselves
2. group media, such as popular theatre, or the "docudrama" employed by the KWDP to explore intra-household responsibilities through role playing (see Rural Case 9)
3. group discussions or community seminars, such as the seminars organized in the Mazvihwa district of Zimbabwe (see Rural Case 6) or the discussions held by the KWDP which stressed socio-economic factors in access to woodfuels
4. the training and visit (T & V) system, as used by CARE in its Agroforestry Programme in Kenya. This is a much more traditional approach, stemming from the idea that community development can be led by specialists teaching lead farmers who will then teach others.

These methods do different things and should be seen as complementary. Which method, or mix of methods, is most appropriate will depend both on local circumstances and priority needs, and on the human resources available.

For instance, the use of group discussions and community theatre by the KWDP enabled a few people to reach a large population and got them to think about solutions to woodfuel problems. In contrast, CARE's T & V approach involved intensive interaction between farmer and technician. Although as effective as KWDP's approach in mobilizing change, it is far more people-intensive and used paid staff with at least secondary-school education. It has most value for providing or increasing awareness of the need for simple inputs, such as nursery facilities. In comparing both approaches, the World Bank's *Kenya Forestry Subsector Review* concluded that T & V would be too costly to employ on a large scale for extension.[19] It has also been criticized for marginalizing some group members and creating dependency.[18]

In fact the emergence of T & V as a development tool nicely illustrates the tensions between Chambers's "professional thinking" and the more holistic approach which we have been advocating here. In their recent insightful analysis of forestry in the context of rural development planning,[1] Harry Blair and Porus Olpadwala point out that T & V was developed to counter some of the defects of the old "community development worker" approach to rural extension of the 1950s and early 1960s, where the extension agent was charged with bringing technological modernity *en bloc* to the countryside. The trouble was that "To master everything he needed to know to promote rural development was obviously an impossible task ... and in even attempting to do so this jack-for-all-trades-but-master-at-none ended up delivering very little in the way of useful knowledge to the villagers assigned to him."[1]

T & V sought to replace the generalist with a specialist who could disseminate detailed technical knowledge on a fixed timetable to "contact farmers", who could then impart the new ideas to their peers. However, for such a scheme to work, standardization and frequent and rigorous monitoring are necessary; this implies the risk of falling into the rigid "project blueprint syndrome", with little feedback *from* farmers. T & V fits all too readily into normal, top–down development professionalism and is thereby both popular and much less effective than it should be.

TOWARDS PROJECT DESIGN

What has all this to do with "the project": the normal vehicle on which governments and donor agencies mount their efforts aimed at rural people? In this section we look at the salient characteristics of tools which might be used for more people-centred (and tree-related) interventions or projects.

A conventional project cycle goes through the sequential phases of data acquisition, analysis, planning, implementation, review and redesign. It is ordered and methodical, but is often a costly and time-consuming exercise. The logical progression is designed to ensure that all factors and considerations are incorporated. But, as many practitioners have shown, such an approach tends to become rigid and bureaucratized in practice. Critical questions are not asked and important insights are missed.

The traditional tools for project design include formal quantitative surveys for baseline studies and later monitoring and evaluation purposes, or ethnographic studies which focus on a small population and rely strongly on participant observation. The serious limitations of these methods include:

- the long time required to produce results
- the high cost of administering formal surveys
- the poor reliability of data, owing to errors and biases introduced by interviewers and questionnaire design
- the irrelevance of many of the questions for specific implementation problems
- the absence of a dialogue with the local community leading to their involvement in planning what is done.

In short, a major shift is required in the methods of gathering information for project design. This is beginning to happen, at least in some places, and in a number of development sectors. The search is on to find more cost-effective techniques which generate findings more quickly and which relate more to the needs and priorities of local people, and to their resources, potentials and capacities.

To outline some key issues which are increasingly seen as being important for good project design:

1. *Explicitly identify target groups.* Unless a commitment has been made to target resource-poor farmers, and this commitment is incorporated in project design, it will only be by chance that they benefit significantly.

2. *Learn about local communities.* The important questions to ask are: what are the resources available locally? What are the extent and nature of local knowledge and existing practices in natural resource management? What incentives exist for local action? What are the nature and extent of local institutions? The key lies in identifying social, economic and cultural constraints and potentials.

3. *Whose knowledge and resources?* As well as understanding each of the groups that need to be involved, the relationships between them must also be considered. Most interventions will involve a combination of outsiders – for example, government agencies, local non-governmental organizations, national or international private voluntary

organizations, and donor agencies – and insiders, the local community. The key linkages between them are those related to the development and transfer of technical advice and knowledge, and those related to resources and their use.

4. *Understand the overall livelihood strategies.* Simply looking at the farming system is not enough. One has to consider the entire production and consumption system, both on and off the farm.

5. *Start small. Have realistic expectations.* It is nearly always wise to start with a pilot project, achieve a measure of success with this, and depend on the demonstration effect to work in diffusing the successful practices or changes at a pace that suits the local community. The time-scale required will often be long, and the intervention will demand a flexible commitment over a long period of time by the outsiders involved.

6. *What information?* All too often project staff take the view that "any information is good information – and the more the better". This is expensive, and frequently wholly unnecessary, not to say misleading. It generates a mass of information with a beguiling appearance of accuracy which is usually far in excess of what is needed or what is good for project design.

This last issue relates to a theme which is central to this discussion of good project design. This theme is "optimal ignorance", a term first used by Ilchman,[20] or "appropriate imprecision".[21] Essentially the argument is that both the amount and the detail of information collected should be the minimum required to formulate useful hypotheses about what an intervention might do. Gathering information uses scarce skills which can often be better used elsewhere. Furthermore:

- more information on more aspects of a community will not necessarily help in designing a better project
- more accurate information will not necessarily improve project design
- all information does not have to be collected anew each time an intervention is planned.

For a more extensive review of social science methods for designing tree-related projects the reader is referred to Molnar, who also includes an annotated bibliography.[22] However, we will consider two types of tools which seek to overcome the limitations of more traditional techniques for project design, and which have been used in the design of tree-related projects.

These tools are agroforestry Diagnosis and Design (D & D) and Rapid Appraisal techniques. The latter is a family of tools with common characteristics, whereas D & D is a more formally defined methodology. In principle there may not be a sharp distinction between them – the designers of diagnosis and design intend it to incorporate a rapid appraisal option – but in practice rapid appraisals tend to be more informal and eclectic.

Agroforestry Diagnosis and Design (D & D)

Agroforestry D & D was developed by ICRAF to diagnose land management problems and design agroforestry solutions. It is aimed at assisting researchers, extension agents, community development workers and NGOs to plan and implement effective research and development projects. A key feature of D & D is its flexibility; it is intentionally open to adaptation to fit the needs and resources of different users, following a pragmatic approach to local problem solving.

The basic procedure follows five main stages, with vital feedbacks between each. This iterative method is central to the D & D concept, in which the experience gained in "learning by doing" is used to reinform earlier stages, and to permit subsequent rediagnosis and redesign. The aim is not to postpone action until a "definitive" plan has been established, but to apply a "best bet" set of actions which can be progressively refined. The five main stages are: prediagnostic, diagnostic, design and evaluation, planning, and implementation. These stages and the all-important feedback loops are represented in Figure 4.1.

The farmers themselves probably undertake the ultimate fine tuning and the dissemination of the agroforestry technologies. Inputs of information from both on-station research and on-farm trials should reinform project design and planning decisions at a later stage, but the project can initially go ahead on the basis of the prediagnostic and diagnostic stages. D & D is designed with a "rapid appraisal" option during the planning stages, with in-depth follow-up during project implementation.

Diagnosis and design is specific to land use systems and not necessarily specific to sites. The results of a D & D iteration may therefore be applied wherever that land use system prevails, and not just

Figure 4.1: Sequence of Activities and Feedback in a D & D-Based Project

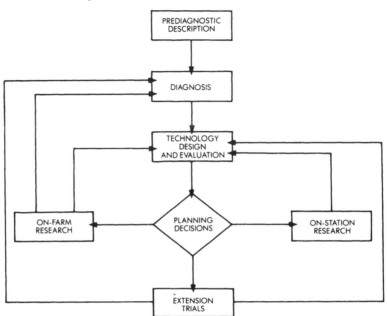

Source: ICRAF (1987).[23]

in one particular place. However, it is stressed that a land use system includes the land users themselves, and so the method may well be site specific. In its ideal form, although one not so far achieved, D & D is a type of farmer participatory research. The iterative nature of D & D is critical in achieving this goal, since creative inputs from farmers may not come until some time after they have been exposed to the experimental technology.

The outcome of the diagnostic procedure is a set of functional specifications for "candidate technologies". A matching exercise follows – matching not only the selected technologies but also their component "nuts and bolts" (species, placing and spacing, management practices, etc.) to a set of differently ranked user needs and demands. It cannot be overemphasized that local people's needs and priorities will not be met by the proposed agroforestry technologies, however, technically feasible they may be, if the people feel – for whatever reason – that they are not for them.

Diagnosis and Design can be used at different scales. It is usually applied either to the basic household management unit (for example, a farm or houshold herd), or at a local community or general landscape scale. However, it is also intended to be applicable in the formulation of co-ordinated national research and extension programmes. At this national policy level, D & D follows similar procedural steps to the smaller-scale applications, but is used selectively to fit the requirements of particular programmes. Adaptations of D & D are being implemented, for example by participating countries in ICRAF's Agroforestry Research Network for Africa (AFRENA).[23]

The main limitations of the household level of analysis are that not all land use problems originate within a single farm, nor can they always be solved by action at the individual household level; and that "the household" is only very rarely a homogeneous interest group. In Africa, men and women within the same household may have very different production rights and responsibilities, and degrees of access to or control over labour, capital inputs and other resources. Their responses to agroforestry options may therefore be very different. A special effort needs to be made to distinguish different types of households and different interests within households. In some areas of Kenya, for example, up to 30 per cent of households are headed by women, and these are often the poorest.[24] The community-level application of D & D is often essential for the proper diagnosis of land use problems based on the differing interests of such groups.

Among the shortcomings of many Diagnosis and Design applications has been the tendency to dwell on the procedural methodology at the expense of a more pragmatic "matching" approach. Perhaps because the methodology appears rather complex and formalized, it has the effect of keeping at arm's length the specific situation with which it tries to deal. The danger – as recognized by ICRAF themselves – is that the D & D user may be inclined to try and "fit" a "pet technology" or preconceived agroforestry "solution" to a situation in which it is inappropriate. For example, ICRAF found that one of their D & D trainees, having applied the procedure "by the book", came up with a proposal to introduce *alley cropping* to Maasai pastoralists in Kenya! As John Raintree acknowledges, "it became painfully clear ... that it was entirely possible to give the *wrong answers to the right questions!*"[25]

The problem with all methodologies which become too rigid in their application is that they can be treated all too easily as *the answer* in themselves. As Diagnosis and Design becomes simplified over time, and as D & D users learn what it can and cannot do, it will hopefully be used more pragmatically.

The Kathama project which is outlined in Rural Case 16 is a good example of what a well-applied D & D can achieve. An important lesson from it is that in designing agroforestry projects, there is no substitute for an awareness on the part of outsiders of actual agroforestry practices, and a feeling for the conditions under which they may be both feasible and acceptable to local people.

Rapid Appraisal Techniques

Most Rapid Appraisal techniques fall under the rubric of "Rapid Rural Appraisal" (RRA) which will be used here as the general term for convenience. RRA may be defined as a systematic but semi-structured activity carried out in the field by a multidisciplinary team and designed to acquire quickly new information on, and hypotheses about, rural life.[26]

Two themes are central to the philosophy of RRA. The first is the pursuit of "optimal ignorance", as discussed above. The second is diversity in information gathering: using several different sources, and means of collecting, information. Information can be verified by the use of diverse sources, in a process of "triangulation", rather than via statistical replication. Secondary data, direct observation in the field, semi-structured interviews, and the preparation and discussion of diagrams, all contribute to a progressively more focused analysis of the situation under investigation.

These themes in turn lead to six key features of good RRAs, namely that they are:[26]

- *iterative:* the process and goals of the study are modified as the team realizes what is or is not relevant
- *innovative:* there is no simple, standardized methodology; techniques are developed for particular situations depending on the skills and knowledge available.
- *interactive:* all team members and disciplines come together in a way that fosters interdisciplinary insights
- *informal:* the emphasis is, in contrast to the formality of other approaches, on partly structured and informal interviews and discussions
- *in the field:* the aim is not just to gather data for later analysis;

learning takes place largely in the field "as you go", or immediately after, in short, intensive workshops

- *involving:* a good RRA encourages a participatory, "bottom–up" approach.

RRA is not the complete answer to all problems in project design. It will never, and was never designed to, make redundant more traditional, formal and detailed surveys and analyses. This is really a question of the scale of the project to be designed. RRAs and RRA techniques essentially complement more formal methods for larger-scale projects. However, on a small scale, such as at the level of a local community, RRAs may be adequate substitutes for conventional surveys. The advent of RRA has widened the scope and choice of methods of analysis for rural development. The important point is that different techniques, both formal and informal, can be blended to suit the nature of the problem, the local resources and the resources to hand.

Agro-ecosystem Analysis

One form of RRA is known as Agro-ecosystem Analysis.[27] This procedure combines a number of different RRA techniques in a structured way to produce an agreed set of priorities for rural development. It shares with other forms of RRA the characteristics listed above, but in particular it:

- focuses on diversity, both biophysical and socio-economic
- reduces complexity to a few key problems and opportunities
- is appropriate to any level of intervention – field, farm, local community, watershed, region, nation
- emphasizes not only productivity but stability, sustainability and equitability, and the critical trade-offs between them.

Two of the most important techniques are semi-structured interviewing and the drawing of diagrams. The interviews may be with individual farmers or with groups, conducted in the field or in the farmer's home. Diagrams may be drawn together with the farmers, and are most useful in identifying specific problems and opportunities which the interviews have revealed. Diagrams are a good way of facilitating communication between specialists from different disciplines.

The RRA team focuses its discussions on these problems and opportunities and then identifies a set of key issues, made up of the most critical problems and/or the most promising opportunities. The team can then decide on a short list of actions – known as "best bets" – which they feel may address the problems and make best use of the opportunities.

Agro-ecosystem analysis was recently used by the Ethiopian Red Cross Society to investigate possible ways of increasing livelihood security in two Peasant Associations (PAs) of Welo region (see Rural Case 24). For one of the PAs, Gobeya, analysis of the seasonal calendar, covering a range of agricultural activities and the incidence of diseases and pests, showed that there was a critical period of the year when a number of problems coincided. Only a multidisciplinary methodology such as Agroecosystem Analysis could have picked this out.

Two of the seven "best bets" suggested to ease the problems identified for Gobeya PA were reafforestation for soil and water conservation and agroforestry for multiple purposes. But, of course, people's felt needs included many things besides forestry development, including drought-resistant, early-maturing crops; small-scale irrigation; home garden development; improved primary health care; and credit facilities. In short, the technique revealed the multiple needs and priorities of local people and was able to help them formulate practical suggestions for action.

Chapter 5

Rural Cases

1: FARMER-LED INITIATIVES IN SHINYANGA, TANZANIA

Shinyanga is one of the most desertified and deforested regions of Tanzania. During the 1920s and 1930s large areas of woodland were cleared during programmes to eradicate tsetse fly and *Quelea quelea* birds; there are signs of soil erosion and land degradation in most areas. Shinyanga has the highest livestock concentration of any region in Tanzania and over-grazing is severe, particularly in the drier Meatu district.

The region is occupied by the agropastoral Sukuma people, who live by a combination of rain-fed agriculture and livestock husbandry. Some 40 per cent of Shinyanga's cultivated area is given over to cotton as a cash crop, which was introduced early this century but with a major expansion in the 1950s. This has increased grazing pressure on the remaining land and pressure on woody vegetation is severe in many districts.

Although there is strong political and party support for afforestation in Tanzania, conventional forestry approaches based on raising seedlings in centralized nurseries are unlikely to succeed, especially in a region like Shinyanga with its severe water shortages, lack of transport to reach the nurseries, and lack of protection for young trees from freely grazing cattle.

However, local people have developed imaginative ways of bypassing this kind of approach and its constraints. For example, farmers have developed ways of propagating trees and shrubs for live fences to control grazing, and have done so using tree cuttings rather than seedlings. Offtake from these live fences is also used for fuel. Direct seeding is practised for a wide range of tree species, some of which are nitrogen-fixing and enhance soil fertility. Some villagers have established thorn hedges to protect saplings.

Local initiatives to manage over-grazing and its sometimes devastating effects on grasslands and woody vegetation have also emerged. In official eyes, land degradation has invariably been associated with overstocking of cattle: destocking polices have been on the official agenda since the 1920s when the colonial authorities first raised the issue. However, since the Sukuma depend upon large herds of cattle as an economic way of spreading risk during drought periods, not to mention their cultural importance, destocking is naturally strongly resisted.

Instead, in a few places community initiatives have been taken to encourage the natural regeneration of woody plants in a more pragmatic way, demonstrating the implicit link made by agropastoralists between the management of both "economic" and biomass resources. As an indication of what such measures can achieve, a 40–50–hectare Forest Division reserve which has been closed to cattle since 1981 has regrown naturally with *Acacia* species at such a phenomenal rate that after five years the canopy was as many metres high.

Traditional strategies include the *Ngitiri* grazing reserves, which are closed to livestock at the beginning of the wet season and opened up, section by section, during the dry season. The idea is to retain an area of standing hay until the rains come. Although this system is not practised everywhere, strict control is maintained over the *Ngitiri* in those villages which do use it. A recent NORAD mission to the region visited Sangilwa village in Kahama district where a violator of the *Ngitiri* rule incurred a fine of 2,000 shillings.

Isagala, another village visited by NORAD, had on its own initiative made a large stretch of *mbuga* land into a village reserve. *Mbuga* lands are moist depressions where rain-fed paddy rice is commonly culti-vated. Isagala's reserve is divided into two sections. One is closed to all forms of use, the other is open to licensed cutting for firewood. Permission to cut is granted to those who have a letter from their party "ten-cell" leader and signed by the village secretary. People from other villages can pay to cut wood. It is possible to implement this scheme effectively because the decisions were taken by the villagers themselves, fully confident that the reserve is *mali yetu* (our property). The reserve is patrolled without payment by the *wasalaama* – traditional guards – of the village.

HASHI: Soil Conservation in Shinyanga

The HASHI programme was established in Shinyanga in July 1986 under the Division of Forestry and Beekeeping in the Ministry of Lands, Natural Resources and Tourism. Since it is involved with soil conservation it has been compared with HADO in Dodoma region, but its style is very different. The programme aims to undertake desertification control efforts which are ecologically sustainable, financially attractive and culturally acceptable, and which can be achieved with locally available resources and capabilities. HASHI's guiding principle is: "Go to the people. Live among them. Learn from them. Start with what they know. Build on what they have."

The main components of HASHI are extension and training, efforts to support sustainable use and conservation of degraded woodlands, reafforestation, agroforestry, soil and water conservation and watershed management, and support to the regional land use planning unit. Its achievement cannot be measured at this early stage, but since the programme is reactive to the initiatives which have already been taken by people in Shinyanga, there are optimistic signs that HASHI will operate sensitively in assisting and enabling local people to mitigate the effects of land degradation.

Sources: ETC (1987),[1] Brandström (1985),[2] Barrow *et al.* (1988).[3]

2: PADDOCKS IN MWENEZI, ZIMBABWE

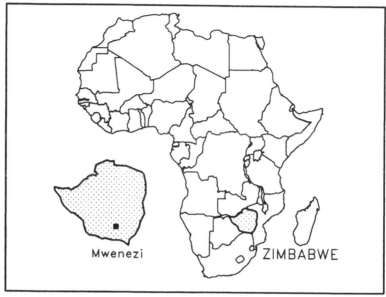

Like the community initiatives in Shinyanga, new indigenous practices in woody plant management have been successful in Mwenezi, Zimbabwe. Success is owed to heightened local perceptions of environmental degradation and its negative welfare impacts, and to the political will of the community.

Mwenezi is a semi-arid area which supports a mixed economy combining semi-intensive livestock husbandry with the growing of drought-resistant grains and maize. Returns to both arable and pastoral production are low and there is a growing vulnerability to drought. Although cattle are vital in the rural livelihood system, over 40 per cent of the population have none and social differentiation within communities is marked. The subsistence farming base is becoming progressively weakened by the migrant wage labour system which takes many men away from the farm to employment in local mines and large agricultural schemes.

In Matibi I, one of the two communal areas in Mwenezi district, the average stocking density is one livestock unit (LU) per 3 hectares, or three times the recommended level under managed conditions of one LU per 10 hectares. There is therefore an urgent need to improve both cattle and pasture management practices.

A new system of land management aimed at doing just this was introduced in 1982 by local initiatives in three wards of Matibi I, each covering approximately 600 households, or 3,000 people. It involves the separation of the household settlement and arable and pastoral areas. In the latter, grazing paddocks are marked out and the rotational "veld management" system, which was first tried in the province in the early 1970s, is adopted. Each paddock is grazed for 14 days, followed by a 56-day fallow period. Seven paddocks in the Mwenezi natural woodland area have also been fenced under the new system, permitting a 21-day rotation for 156 cattle. Following the crop harvest, cattle are released to feed on the cereal stalks left in the fields, while their manure benefits soil fertility and structure.

Live fences are used to mark off some of the paddocks to provide extra fodder, construction poles, fuelwood and other products. Orchards of macadamia and cashew trees have been introduced to provide cash earnings and fuelwood offtake. Gully control measures are under way in the district, including the planting of sisal which can also provide fuelwood.

Outside donor support is helping to build on these grassroots initiatives. For example, the wire fencing which is often used for paddocking is costly and made possible only by an EEC grant. In time, it is hoped to replace it with the maturing live fences so that the same wire can be reused to fence new paddocks. The challenge for outsiders who become involved with local initiatives of this kind is to lend support sensitively in a way which will not swamp the very initiatives they are trying to support. Simply pumping in money will not help.

The short-term benefits of these community initiatives are already apparent. The average price of cattle has risen from Z$100–150 to Z$400–450, giving higher incentives to farmers to market cattle. This price increase partly reflects improved quality, but also the important fact that the way in which people value their livestock has changed. Cattle are now being seen not so much in terms of their social value but as "cash on the hoof". Even more significant will be the long-term benefits in terms of woody plant resources and enhanced sustainability of the rural livelihood system.

Source: ETC (1987).[1]

3: BUILDING ON INDIGENOUS PRACTICES IN TURKANA, KENYA

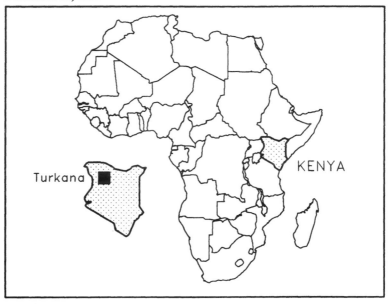

The Turkana are a pastoral people with a rich base of traditional knowledge about their environment. They live in the driest part of East Africa, where annual rainfall varies from 400 mm down to only 150 mm, well below the normal limit for rain-fed agriculture. The Turkana combine a nomadic life-style, moving in search of good grazing in the dry season, with sedentary occupations, according to varying local circumstances. Despite the arid conditions, agricultural cropping is still practised. As an adaptation to drought conditions, the Turkana have themselves developed a sorghum which is the fastest-maturing variety known, requiring just sixty days in the ground from planting to harvesting.

The rivers in the area are seasonal and the major ones are lined by forests. These are the main tree resources in the district. Along the Turkwel River, for example, the riverine forests vary in width from one kilometre or so in the north to 5–10 kilometres further south. Although pastoral groups are normally associated with communal land tenure, much of the riverine forest is divided into individually owned plots known as *ekwar*. These have major economic and cultural significance for the Turkana, which outsiders are only just beginning to appreciate.

They produce fruits, construction material and medicines, and are an extremely important source of dry season fodder and browse for livestock. The dominant tree species is *Acacia tortilis*, but other important species include *doum* palm (*Hyphaene coriacea*) and *Cordia sinensis*. Apart from their production role, many of the *ekwar* plots are the site of family graves, and their owners often feel a strong personal attachment to them.

The *ekwar* system has fallen somewhat into disuse, in part because of the dislocations caused by the recent drought and by outside influences. For example, a new tarmac road connecting this remote district with urban centres in the rest of Kenya has led to greatly increased cutting of the *Acacia tortilis* trees for charcoal in some of the riverine forests. Areas of the forests are becoming degraded and wherever the population pressure is high, no regeneration of the *Acacia tortilis* is taking place.

Turkana Rural Development Programme (TRDP)

The forestry component of the Norwegian-funded TRDP, which operates through and gives institutional support to the District Forestry Office, has been trying to revive the traditional system of ownership and management of natural vegetation, particularly of the *ekwar* plots. TRDP is now trying to integrate this and other traditional systems of resource use (such as traditional grazing reserves) into current Government of Kenya policies and regulations regarding such matters as charcoal production and trade, timber-felling permits and forest clearance for agricultural purposes.

The TRDP has carried out much field work to document these traditional systems. It has also conducted courses and seminars which have brought together elders, chiefs, other local leaders and foresters from the Kenya Forestry Department's Rural Afforestation and Extension Service. The main aims of these educational sessions – which are a learning experience for all participants – have been to reconcile traditional and modern administrative systems, as well as to define management approaches to local vegetation based on both traditional knowledge and modern forestry.

This work of bringing together the traditional use of vegetation and contemporary forestry is not easy. It requires infinite patience and much political and anthropological skill. For one thing, local knowledge of the vegetation and its management, which has accumulated over the centuries, is much too valuable not to retrieve and use. It would be foolish to overlook local systems of resource allocation and use because to do so is to risk alienating the very people whom outsiders are trying to serve.

The TRDP has also experimented with exclosures. These are simply sections of over-exploited vegetation which have been fenced off to allow the vegetation to recover on its own, but with the agreement, co-operation and understanding of the local people. Initial results, in terms of recovery from both seeding and vegetative regrowth, are encouraging, although the natural mortality of seedlings in such a dry area can be very high.

A most encouraging aspect of these exclosures is that fencing is not essential. The real test of extension forestry in dry regions is to get local people to agree not to use certain areas until the vegetation has recovered adequately, without fencing or strict policing. In Turkana district it seems that this is starting to be achieved. People are voluntarily protecting and encouraging the growth of young *Acacia* and *Dobera glabra* trees, especially around their homes. In some places where mature trees are all of the same age class (indicating that regeneration has all but ceased), young trees can be seen today.

Microcatchments

The TRDP has also been trying out various forms of rainwater harvesting to help afforestation efforts. Without some way of concentrating rainwater runoff from a relatively large area into a smaller area where the tree can be planted, tree establishment in such a dry environment as Turkana would require prolonged and time-consuming watering.

Collaboration between OXFAM and the District Forestry Department led to trials using V-shaped microcatchments of different sizes with a range of tree species. The microcatchments are all hand-dug, using only simple, locally available tools. The standard size is 10 by 10 metres, which when full contains approximately 3,000 litres of rainwater, and takes one person one day to construct. Several different tree species have been successfully established with microcatchments of this size.

However, trials were also conducted with much smaller catchments of only 5 by 5 metres, which are normally used only in wetter areas. Although tree establishment is naturally more difficult with these smaller units, one person can make four of them in a day. The point of the trials is to see whether, in terms of labour investment per unit area, the smaller microcatchments generate a higher production of woody biomass than the larger ones.

Alongside these trials, the TRDP is facilitating the establishment of trees by local people, using the same microcatchment techniques, but relying wholly on social agreement as a way of protecting the young trees from browsing animals. Local people are paid for the construction of the catchments by food-for-work disbursed through the Turkana Rehabilitation Project, with the district forestry department providing technical advice.

Quite staggering growth of *Prosopis* trees can be seen in these planting areas after only one year. People feel that the trees are their own and make great efforts to keep their livestock away from the planting sites. Not only do the microcatchments provide a viable means for establishing trees, but also visibly serve to improve the growth of ground vegetation. During the second year of establishment, even crops such as drought-tolerant sorghum and/or cowpeas can be grown.

Sources: Barrow (1986, 1987a, 1987b),[4] Kenyan Forestry Department, Extension and Information Service (1988),[5] Mearns (1988).[6]

4: THE KOUMPENTOUM ENTENTE, SENEGAL

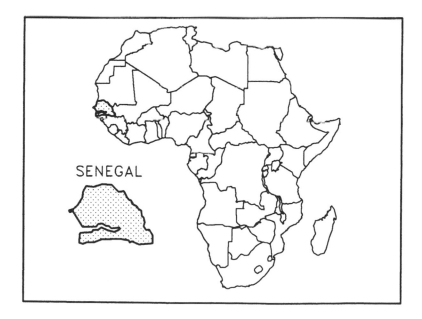

The Koumpentoum Entente is an association of self-help village development organizations in Senegal. It was founded in 1974 after the first wave of modern Sahelian droughts, by an exceptional group of villagers concerned to do something about their environment. The small nucleus of the organization received help and encouragement from Mahamadou Cissoko, a famous Senegalese extension worker. After more than ten years of experience, virtually every activity is planned by the villagers themselves before they seek outside financing. Participation is voluntary and some activities are self-financed.

The village of Diam Diam, for example, has undertaken a series of natural resource management initiatives. In 1981 the village subcommittee set aside 250 hectares of village land for reafforestation. Eight hectares of fenced plantations were set aside for Eucalypts, Australian Acacias and indigenous species; of these 7 hectares have been planted. Five hectares have been fenced for a future village orchard. A woodlot and windbreaks have been established as demonstrations. The youth group and the women's group have planted 50 hectares of eucalyptus each. The village nurseryman has gathered seeds and will begin to regenerate lost or scarce indigenous tree species. In addition, 6 hectares of fields per year are fertilized through a planned, staggered programme of livestock corrals for collectively managed but individually owned herds.

In addition to these natural resource initiatives, Diam Diam has built a collective granary, a co-operative store, a small, unstaffed health clinic and a warehouse, and is stocking bricks to build a mosque. The Entente purchased a millet thresher for the village, and the village association itself purchased a millet mill. A clever scheme is also under way by which co-owned goats and cattle are used to create the nucleus of individual herds. The village subcommittee expects to receive new technical assistance to launch an improved stove production programme in the village.

The administrative structure of the Entente centres on the village subcommittees which oversee each project, with a single person responsible for each of the projects. Funds were originally generated from collective fields and this mechanism is still used. However, the Entente has received outside support from at least half a dozen NGOs and other donors, the major one of which is Catholic Relief Services, who have provided technical assistance and financial backing for over ten years. Each Entente project searches for separate technical and financial assistance, if needed, either from an NGO or a government service. The initiative, however, rests squarely with the villagers themselves.

Source: Shaikh *et al.* (1988).[7]

5: LEARNING TOGETHER: FORESTRY DEVELOPMENTS IN KONSO, ETHIOPIA

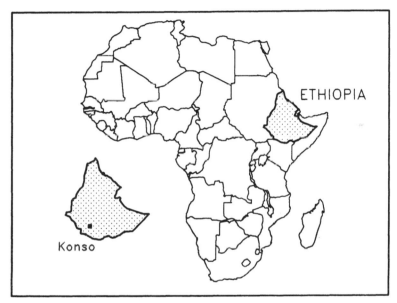

One successful programme involving tree planting in semi-arid agro-pastoral areas is the NORAGRIC-supported Konso Development Programme (KDP) in south Ethiopia. Tree planting began in 1977 as part of an integrated community development programme aimed at reducing soil erosion and providing fuelwood and building poles. The key difference from similar programmes in Ethiopia is that due to KDP's flexible "learning" approach, other objectives including the improvement of soil fertility and animal fodder supplies were added to the programme, in line with the expressed needs and priorities of local farmers and herders.

Altitudes in Konso range from 1,300 to 1,800 metres, while annual rainfall in the upland areas where most people live is an unreliable 600 to 800 mm. Crop yields in recent years have been poor due to particularly scarce and unpredictable rains and exhausted soils.

Population density averages fifty-five persons per square kilometre. An acute fuelwood shortage is starting to be felt near villages and most households also use crop residues (the stalks of sorghum, maize, cotton and pigeon pea) as fuel.

Tree Planting in Konso: The Learning Curve

The KDP began by using food-for-work incentives to get local people to plant trees, and relied on seedlings from state-owned nurseries some distance away. Planting was restricted to eroded, dry communal land that had been rejected for farming, to reduce the likelihood of land use conflicts.

As confidence gradually built up, planting started to take place in better locations, although not without problems. Where there was a shortage of land, for example, farmers began cultivating crops in previously planted areas, demonstrating that the value to them of the food payment exceeded the future value of trees. In some cases the land allocated for tree planting by Peasant Associations (PAs) had been individually owned and "unintended" fires broke out as a form of protest.

Successful tree planting in dry areas is always difficult, especially where there are browsing animals to contend with. After the first few years, full responsibility for tree planting in Konso was assumed by the local PAs, who now propose all planting sites. The PAs are committed to the protection of tree seedlings and restrict grazing by means of fencing with thorny bushes. There are now seldom conflicts over grazing in planting areas. Local people are personally involved to the extent that there is purposeful culture of local bushes, or even planting along paths and field boundaries to protect soils and crops.

A similar process of "handing over" has occurred with planting materials. Reliance on seedlings from distant state-owned nurseries soon gave way to the production of seedlings in KDP nurseries close to the planting areas. Seedlings could then be carried by hand rather than by vehicle, a practice which will be far more sustainable after funding for the programme is withdrawn. But in Konso and elsewhere it is now received wisdom that with the right seeds, equipment and training, local farmers could run "decentralized" nurseries producing a range of multi-purpose tree seedlings.

Agroforestry

Indigenous agroforestry practices have been a feature of Konso rural livelihoods for a long time. Self-sown multi-purpose tree species are tended in cultivated fields and some are deliberately planted. The latter include the drought-resistant *Moringa oleiofera*, the leaves of which are cooked in combination with "sorghum balls"; and *Terminalis brownii*, which coppices easily and has leafy branches which are lopped for animal fodder, especially for stall-fed sheep. These species are considered inferior for firewood, however. *Acacia* species are not normally planted, and their self-propagation is considered sufficient by the Konsoes. The most often-requested species for private planting is *Eucalyptus globulus*, now widespread in more humid parts of the Ethiopian highlands following its introduction at the end of the last century. However, it is successful only in the best locations, for which there may be competition with higher-priority food crops. In the drier areas of Konso, *E. camaldulensis* is preferred for poles and fuelwood, but is even less suited to local conditions than many local species.

The KDP has learnt from experiences like that of the Borkana Catchment project, where the low success of reafforestation efforts was due in large measure to the inappropriate use of Eucalypts, and has shifted its focus now to more ecologically appropriate indigenous species. Such species are also more socially appropriate, since they allow for participatory research and development, drawing on the Konsoes' own considerable wealth of local knowledge about tree species and their functions.

Source: Haugen (1987).[8]

6: RESEARCH IN THE MAZVIHWA, ZIMBABWE

A study which takes indigenous technical knowledge as a point of departure was carried out in the Mazvihwa and surrounding areas of Zimbabwe for ENDA-Zimbabwe,[9] prior to the launching of a community-based tree management and extension project towards the end of 1987. This work is notable in that a very extensive body of indigenous ecological knowledge has been drawn upon and systematically recorded in the form of:

- a vernacular tree dictionary, matching local tree names with their botanical species names, as far as possible

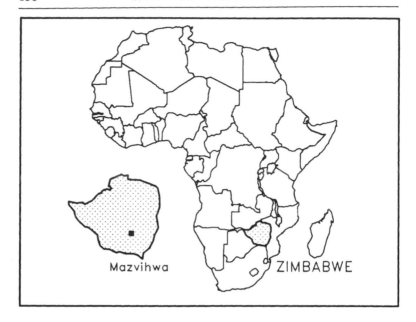

Mazvihwa   ZIMBABWE

- an inventory of tree resources in the area, giving comprehensive information for a wide range of indigenous tree species on: ecological conditions; wood quality; end-uses of offtake; effects on crops; and reproductive ecology
- additional information on the ecology of deforestation in the region, and tree tenure rules, as these provide the context for research into stresses on and uses for tree resources, as perceived by local people.

As part of the research project, several community seminars were organized to provide opportunities for discussion – often long and heated – of problems and potentials in woody plant management with local people. It is they who will decide ultimately which tree species are most appropriate for which end-use needs.

The ENDA woodland management and tree extension project which was formulated as a result of this study aims to help local communities, at Ward Development Committee and Village Development Committee levels, to draw up woody plant management plans based on the rapid appraisal of their indigenous technical knowledge base. The project employs an approach to community planning which is designed to be run by farmers themselves.

Woodland management options being considered include control over cutting patterns, tree planting and protection in communal grazing areas, trees around homes and in fields, and live fencing. Community workers who are also local farmers will receive a short training in nursery and tree production techniques and basic ideas in community resource management. They in turn will train farmers as necessary. ENDA provides administrative back-up and co-ordinates implementation.

In discussion with people in the Mazvihwa it became apparent that while they had faith in the traditional ways of managing resources, they had recognized that new problems and stresses had emerged which were beyond the coping capacity of the old mechanisms. "Individuals and communities reacted to the idea of the woodland management and tree extension project by saying that it was so obvious, and so obviously within their grasp and knowledge, that they wondered why they were not doing it already!"[9]

Sources: Wilson (1987),[9] Scoones (1988).[10]

7: CHITEMENE SHIFTING CULTIVATION, ZAMBIA

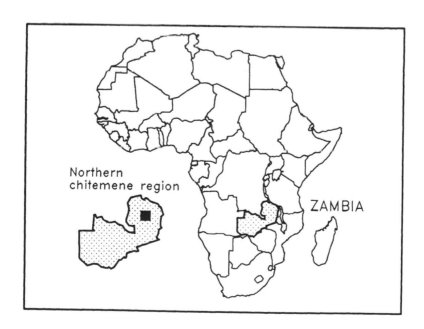

Northern chitemene region

ZAMBIA

Research into the *chitemene* system of shifting cultivation in the miombo woodland of north-east Zambia provides a good example of how agroforestry research can be guided by indigenous technical knowledge.

Shifting cultivation systems are sustainable in areas of low population density where there are only weak links to commercial production systems and markets. With increasing populations and food requirements, and greater market penetration, there is a need to develop alternative cultivation systems which are more sustainable (ecologically, economically and socially).

Of the many types of shifting cultivation practised in Zambia, *chitemene* is unique in that crops are grown in an ash garden made from the burning of a pile of branches obtained by lopping and chopping trees from an area many times larger than the garden itself. One study in northern Zambia found that the average size of plot in which the crops are grown was 0.25 hectares, while the average surrounding area from which the wood was cleared was 2.4 hectares. This system is especially suited to the leached and acid soils of northern Zambia because it enhances the yields of finger millet. The burning suppresses harmful microbacterial activity in the soil and allows a steady and continual release of nitrogen-rich ammonia throughout the growing season.

A study carried out by Professor Chidumayo[13] in three districts of Zambia demonstrated how the *chitemene* system persists in spite of a growing rural population, largely because fallow periods have been shortened. With progressive deforestation, fallow periods have been halved from around twenty-five years to twelve years or so. The frequency of clearing new *chitemene* gardens has been reduced from once yearly to once in two years. The decline in woodland in parts of the *chitemene* region seems to have led to the introduction of many small ash circles in a larger, often clear-felled, area to compensate for the scarcer woody biomass. Chidumayo estimates that, very roughly, these responses have artificially increased the density of population which can be supported from around two to eighteen persons per square kilometre. The area of woodland cleared per person has been halved, from about 1 to $\frac{1}{2}$ hectare.

Research into more permanent cultivation systems is being carried out by the Soil Productivity Research Programme (SPRP) of the Zambian Ministry of Agriculture and Water Development. An important component of this programme has been research into improved fallow systems in north-east Zambia undertaken at the Misamfu Regional Research Station.

An informal survey of land users near Misamfu was carried out by a team from the Ministry of Agriculture, ICRAF and NORAD. This revealed a wealth of information and opportunities for collaborative experiments on innovations and areas of research which were initiated and defined by local farmers. These included:

- a mounding system using grass "compost" to improve soil fertility
- tree culture in women's home gardens, developing towards multi-storey food production systems
- tree domestication in hedgerows
- management of miombo woodland for increased production.

Farmer-led experimentation into compost mounding gave rise to the *fundikila* cultivation system, a semi-permanent system with an estimated population-carrying capacity of 20–40 persons per square kilometre compared with only 2–4 persons for the *chitemene* systems. However, owing to rather better soils in the areas where the *fundikila* systems were studied, the actual increase in carrying capacity from adopting these new methods is less than this tenfold difference, though still large.

Wealthier residents of the area are intensifying their agricultural production by adopting hybrid maize and fertilizers, but this option is not open to resource-poor farmers. They, on the other hand, are actively engaged in experimentation with mounding systems and tree culture. Women are involved in these experiments at least as much as men. Farmers have not yet fully incorporated trees into the *fundikila* mounding system, but the survey found them to be eager to experiment with trees which would produce high-nutrient leafy biomass for use as mulch in the mounding process instead of grass, which is relatively scarce in many areas. They also readily discussed how to incorporate trees into mounded plots, fallows and the local miombo woodlands.

Chitemene cultivation has been one of a series of applications of the agroforestry methodology developed by ICRAF, which is discussed in Chapter 4. It begins with a rapid diagnostic survey to identify potential roles for specific agroforestry interventions in removing farm-level production constraints. The methodology need not be restricted to the farm scale, however, and in the *chitemene* case it is applied at a larger-than-farm scale. The introduction of local knowledge into a research plan for agroforestry in this area or Zambia, including improvements suggested by the D & D survey, is summarized in Table 5.1.

Table 5.1: A Research Plan for Agroforestry in the "Chitemene" system of North-east Zambia Incorporating Local Knowledge

PREVIOUS RESEARCH DIRECTIONS
Improvements to *chitemene*
 • Modified burning and improved fallow methods
Agroforestry alternatives to *chitemene*
 • Alley cropping of fertilized, limed, maize with three exotic tree species: *Leucaena leucocephela, Gliricidia sepium*, and *Calliandra calothyrsus*

RESEARCH SUGGESTED BY DIAGNOSIS AND DESIGN SURVEY
Modification of *chitemene*
 • Changes in spacing/geometry of plots
 • Enrichment of plot boundaries and nearby miombo with more careful choice of indigenous and exotic multipurpose tree species
 • Intensify/improve intercropping practices
 • Soil conservation techniques
 • Techniques to prevent leaching of nutrients
 • Cover crop and tree mixtures to plant at end of cropping cycle

Modification of mound systems
 • Study size and spacing relative to plant residues, cropping cycle
 • Boundary or alley cropping with fast-growing woody species to produce plant residues for mounds
 • Study decomposition/nutrient cycles for grass and woody plant residues

Plant propagation trials

Home garden intensification and diversification
 • Integration of higher-value fruit trees (commercial or home use)
 • Incorporate boundary or alley plantings of multipurpose trees to improve crop yields and maintain permanent cultivation on plots
 • Diversify women's cassava plots ecologically and economically
 • Introduction and trials of leafy vegetable crops

Enrichment of miombo woodland
 • Replace existing species with more productive ones
 • Increase the proportion of useful, preferred species in fallows and miombo areas near homesteads

Screening of productive trees for dambos (seasonally flooded depressions)
 • Screen multipurpose trees for tolerance of waterlogging and soils
 • Test multipurpose trees for production on dambo boundaries of residues for mounded vegetable plots

Source: Rocheleau (1987).[11]

Sources: Rocheleau (1987),[11] Singh *et al.* (1987),[12] Chidumayo (1987).[13]

8: ALLEY FARMING AND DAIRY DEVELOPMENT IN KENYA

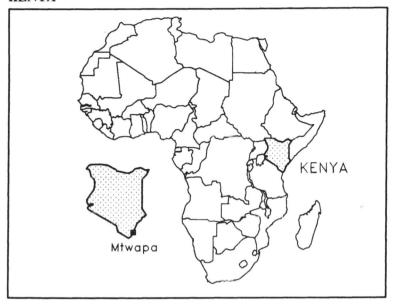

Six Agroforestry/Energy Centres have been established in Kenya under the Ministry of Energy and Regional Development and funded through the Kenya Renewable Energy Development Project (KREDP). A major component of KREDP is its agroforestry programme.

The Mtwapa Agroforestry/Energy Centre was set up in 1982 to develop an agroforestry solution to the problem of low and declining soil fertility in the coastal Kilifi district. The problem was defined by agricultural researchers at the Coast Agricultural Research Station, who proposed the solution of planting food crops in parallel lines or alleys with leguminous trees and shrubs, particularly *Leucaena leucocephela*. The suggested technical package was developed on-station and targeted at farms in the high potential agricultural lands which make up 8 per cent of the district's area.

On leaving the research station, however, the alley-cropping package was modified substantially by farmers into an alley-*farming* system in which the *Leucaena* trees were valued more as a source of animal leaf fodder than as a way of increasing soil fertility and food crop yields. Although the original technical package did include fodder production, the potential of this component had not been fully realized.

The benefits to farmers were felt in terms of higher cash incomes from the sale of milk and of farmyard manure for food crops, especially vegetables. The small decline in maize yields from the space taken up by the inter-planted trees was offset by these benefits. *Leucaena* fodder was able to bridge the dry season fodder gap,[14] while the market value of improvements in herd quality and milk production exceeded that of any possible equivalent increase in maize yields. Following this modification of the agroforestry package, which was at least partly farmer-led, the innovation was rapidly taken up by farmers in Kilifi district.

The case clearly demonstrates how flawed problem definition can lead to inappropriate project design. This can be avoided if indigenous technical knowledge and people's priorities are solicited at the outset. It also emphasizes the research and development capabilities of local farmers in modifying an agroforestry package to suit their own needs.

Another project which worked with the same farmers was the Dairy Development Project, under the Ministry of Agriculture and Livestock Development. Despite their complementary aims, organizational factors ensured that this and the alley-cropping work were not co-ordinated and, indeed, conflicted to some extent. It fell to the local farmers to take an active lead in identifying the missing links. The use of *Leucaena* fodder in the alley-farming system ensured that cattle sold through the Dairy Development Project survived the dry season with adequate supplies of fodder. Without this farmer-led initiative it is unlikely that they would have done.

Source: Jama (1987).[15]

9: THE KENYA WOODFUEL DEVELOPMENT PROGRAMME

The Kenya Woodfuel Development Programme (KWDP) was set up in 1983 following the Kenya Fuelwood Cycle Study, carried out by the Beijer Institute at the invitation of the Kenyan Ministry of Energy and Regional Development. The most important conclusion of the study was that while the wood resources of the sparsely populated semi-arid regions of Kenya are effectively being "mined" to meet predominantly urban woodfuel needs, the greatest fuelwood shortages are in fact being felt in the densely populated areas of the country with high agricultural potential.

In these areas pressure on the natural bush has become severe owing

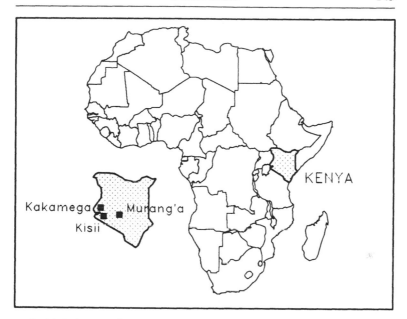

to the high degree of land demarcation and privatization. This is despite the often large numbers of trees and shrubs growing on the farms which should have contributed significantly to the supply of fuelwood but obviously did not. The KWDP's first task was to find out how the trees on farms came to be there, and since they tend not to be used to supply fuelwood, what they are used for, and by whom.

The KWDP currently operates in two high potential districts (Kakamega and Kisii) and has completed resource studies in a third (Murang'a). Population densities range from 260 to 900 persons per square kilometre. Work began in Kakamega in 1984 with the original intention of establishing self-sustaining tree planting systems to improve the supply of woodfuel at the individual farm level. It was extended to Kisii and Murang'a in 1985.

The initial objectives of the programme were:

- to build on existing skills
- to help local groups establish and maintain nurseries to produce seedlings for their farms
- to upgrade existing nurseries
- to run an extension programme to improve nursery- and tree-management skills.

These objectives and the understanding of the problem were transformed by a reconnaissance survey of 528 households in Kakamega district and later studies in all three districts which revealed that woody biomass resources were more extensive than previously assumed. In parts of southern Kakamega, where population density is extremely high (with pockets of up to 1,500 persons per square kilometre), more than 20 per cent of the total land area is covered by trees and shrubs, many of them deliberately planted. Indeed, in Kakamega as a whole, the area covered by farm trees and shrubs increases as the population density increases and as average farm size decreases. Over 75 per cent of the trees on the farms have been planted by farmers themselves.

The observed fuelwood shortage in Kakamega was therefore not the result of a physical scarcity of wood resources on individual farms, or of a lack of local expertise in tree growing. Rather "it is due to social and cultural forces within households which determine control over and access to the wood produced on the farms".[16]

In response to these findings, the programme began to examine the motivation for tree planting and rights of access, through discussions with men and women, often separately. What they found has been summarized as follows:[17]

> Men have overall control of household resources, permanent ownership rights in land and exclusive rights to plant trees. Women have rights of access to household land but not to the trees. Women find it increasingly difficult to fulfil their responsibility for collecting firewood from common land. They are therefore forced to reduce household energy consumption and intensify their income generating activities to earn cash for fuelwood purchase.

The KWDP currently expects that these gender differences can be bridged by a community extension programme which will create an awareness of woodfuels as a family problem, hopefully leading to systems of tree planting, management and exploitation by both men and women.

Mary Kekhovole, one of KWDP's extension workers in Kakamega, describes their extension method as one of "holding up a mirror" to people's cultural practices in tree and shrub management so that they can see for themselves the shortcomings of the present gender divisions. People are seeking solutions for themselves, too, using the tree seeds provided by KWDP. The present extension strategy is to promote tree species which are relatively fast-growing and do not yet have a cash value attached to them, but could be planted for fuelwood production.

For example, one problem the KWDP faced in parts of Kakamega was to find a way of changing attitudes towards the planting of trees by women on household plots so that this would not be perceived by men as a challenge to their authority. A possible solution was found in a shrubby tree which is not categorized as a tree with respect to rights of use or the usual taboos against women planting. This is *Sesbania sesban*, which is traditionally interplanted by women with food crops to increase soil fertility. The KWDP saw the potential in developing this and other tree species, especially *Calliandra calothyrsus*, for use by women as fuel.

The mass awareness programme organized by KWDP has taken an open-ended approach which has greatly facilitated farmers' free experimentation and expression of opinions. The catchword of the programme is *Moto mwaka*, "firewood all the year round". Its methods include popular theatre, known as the "docudrama", in which men's and women's roles are played out in an exploratory fashion and can be openly discussed afterwards.

If the KWDP can continue to develop effective and practical solutions to the problem of rural fuelwood supply it intends to hand them over to other institutions, such as the extension services of the Ministry of Agriculture and Livestock Development and the Forestry Department; schools; or non-governmental organizations that work in and with rural communities.

Sources: Chavangi et al. (1985),[16] Farrington and Martin (1987),[17] Chavangi (1984),[18] Bradley (1988),[19] Ngugi (1988).[20]

10: WATER HARVESTING IN YATENGA, BURKINA FASO

In 1979 OXFAM began an agroforestry project in Yatenga, in the north-west of Burkina Faso. This region in the southern Sahelian zone, one of the most densely populated areas in the country, had been misused for years and has spreading areas of heavily crusted and nearly impermeable soils. The average population density in 1976 was 141 persons per square kilometre of agricultural land and in many villages 50–75 per cent of the land was under permanent cultivation. Crop yields had fallen to disastrously low levels and farmers were extremely vulnerable to drought.

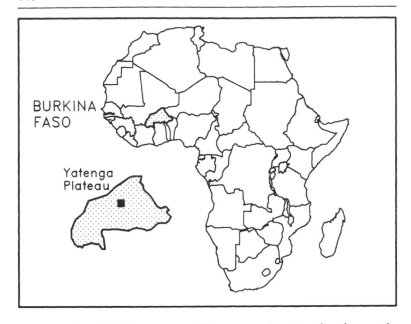

The project initially encouraged farmers to plant trees in microcatchments, a simple water-harvesting technique designed to capture available water to improve seedling survival and tree growth. This idea was taken from a similar system in the Negev, Israel, which one of the OXFAM staff had visited. However, the farmers in Yatenga soon noted that water harvesting had a remarkable effect on crop yields, which to them was more interesting than any effects on trees.

The farmers also identified the microcatchments with a traditional technique of "soil pitting" which they practised already and which involves digging small depressions in the soil about 0.2 metres in diameter spaced about 1 metre apart. The pits capture rainwater runoff and concentrate it in the root zone of plants sown in the pits. Manure may be placed in the bottom of the pits, which also trap wind-blown leaves as mulch. Termites come to feed on this and leave tunnels behind them which facilitate root growth and increase water infiltration. The termites have usually gone before the crop is sown and have not been observed to do any harm.

Farmers extended this practice by sowing cereal crops in the microcatchments when they saw that with this treatment even some of their previously abandoned fields could outproduce their good fields, particularly in drought years. They recognized that runoff farming is an effective strategy for minimizing risk, since its main benefit is to improve the reliability of yields in drought years. This effect is seen in Table 5.2 where the highest yield increase occurs in the year of least rainfall (1984).

Table 5.2: Evolution of Annual Rainfall and Cereal Yields in Yatenga

year	annual rainfall (mm)	control plots		treated plots		Increase in yield (per cent)
		number	yield (kg/100m²)	number	yield (kg/100m²)	
1981	692	3	5.10	14	8.57	68
1982	421	45	4.42	47	4.95	12
1983	413	37	2.95	63	4.18	42
1984	383	72	1.53	74	2.92	91

Source: Reij (1987).[21]

After three years the project responded to further farmer-led changes. First, the farmers abandoned the labour-intensive microcatchments in favour of another traditional technique: contour bunds, or low barriers running along contour lines and designed to trap water runoff. Second, instead of the traditional short earth barriers, the farmers and the project jointly developed a technique based on extended stone contour bunds. Unlike the conventional bunds, the stone lines are permeable and therefore have a dual purpose: they can not only harvest available runoff in drought years, but can also control excess runoff and prevent sheet erosion when rainfall is ample. The bunds slow the runoff and allow more rainfall to percolate into the soil.

Constructing the bunds, however, presented a problem. The land was quite flat, with only a 2 per cent slope, so that it is difficult to determine the contour. An improperly laid contour line will channel water to its lowest point and can cause a gully to form. Previous rock bund construction had depended on government agents with levelling equipment, a serious constraint to the spread of the technique. To overcome this the project introduced a simple levelling device costing only around US$6. Called a water tube level, it is a clear garden hose filled with water which can be moved up and down the slope until the water levels at each end are at the same height, thus determining the contour. Farmers can learn how to use the level in two to three days and can then teach their neighbours. During the second phase of the project (1983–6) about 1,750 farmers in 339 villages had been trained in its use and in rock bund construction. Participants in the training course are selected by their fellow villagers, with three to five trainees per village.

The Yatenga training programme aims to teach a single crucial but very simple technique to a large number of farmers. This may be of more value than a programme which promotes many ideas simultaneously but reaches only a few individuals. It also allows the farmers' own insights and techniques to be more fully used. By trial and error farmers themselves develop the most appropriate methods. For example, if to save labour the barriers are spaced too far apart to be "optimal" for crop production, they still confer some benefits, especially erosion control. Farmers can then adjust the spacing using their own economic assessments and priorities – and can do so better than anyone else. As one farmer said of his newly constructed terraces, "If I see during this season that they are too far apart, I'll add some new ones in between them next year."

As well as the construction of stone bunds, grassed terrace borders are also encouraged. Bundles of crop residues (sorghum and millet stalks) are staked down on the contour where the terrace is to be formed and perennial grasses are implanted or sown upslope. This barrier retains moisture and gives shelter to the grass, and if protected from livestock, both crop stalks and grass will promote a gradual build-up of the terrace.

Although relatively wealthy farmers have taken the lead in rock bund construction, the low capital investment has meant that resource-poor farmers have also taken it up. However, the collective treatment by men and/or women of communal fields proved less successful than the treatment of private fields. This is because farmers first devote all their labour and available organic manure to their own fields to assure their own food security and only then do they tend communal fields, which are inclined to be on the poorest soils anyway.

However, the project introduced another innovation to try to surmount this problem, by encouraging the collective treatment of individual fields. Normally this would require farmers to remunerate the work party with food and drink. As resource-poor farmers could not afford to do this, the project worked with the more organized communities to set up a scheme of maize flour loans which individual farmers could draw on to prepare food for the working parties. A revolving stock of maize flour is maintained, and loans are reimbursed after the harvest. In over half the villages during 1984 to 1986 the original maize flour stocks were reconstituted or better. This incentive to farmers to engage in rock bund construction has proved successful, certainly more so than food aid as used in other soil and water conservation projects in Burkina Faso. It is more sustainable, as it places more control in the hands of local communities themselves, and the increases in crop yields provide sufficient short-term benefits for the continued participation of the farmers.

Achievements and Further Adaptations

With the collaboration of the Ministry of Agriculture and Livestock, the rate of rock bund construction has doubled each year since 1983. By 1986 some 2,500 hectares had been treated, while crop yields have increased by between 40 and 180 per cent. Yet the total cost of OXFAM's investment has been very low: less than US$150,000 during the first six years.

The success of the Yatenga project is attributed to its joint efforts with farmers to develop their traditional practices and to its effectiveness in meeting an urgently felt need. The contour barriers offer an immediate increase in crop yields and reduction of drought risk. However, the project and farmers' strategies are still developing. The project leader has recently expressed "guarded optimism" that the food security gained by the farmers from the first treated fields will enable them to concentrate on the longer-term problems of soil fertility in later efforts; for example, by planting trees. The planting of leguminous acacias on crop land had been a traditional practice, and some farmers have already begun to plant trees along their contour bunds, as well as encouraging the natural regeneration of trees and bushes where the barriers have stabilized the soil.

Sources: Reij (1987),[21] Kramer (1987),[22] Pacey and Cullis (1986),[23] Twose (1984),[24] Wright and Bonkoungou (1985).[25]

11: COMMUNITY FORESTRY IN NORTHERN SUDAN

Several community forestry projects have been established in northern Sudan under the government- and USAID supported Sudan Renewable Energy Project (SREP), with technical support from the Sudanese Energy Research Council (ERC). The experience of these projects to date shows that not only is people's participation an important component of success, but that it can be the poorest and most remote communities who have the greatest resources of organization and enthusiasm to bring to forestry. Furthermore, the best results are often

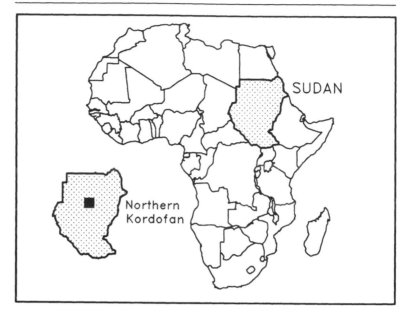

achieved when maximum responsibility for project development and management is given to the community.

The Um Inderaba Village Project.

The most successful project was one in Um Inderaba village where, by 1983, local concern for the environment was strong for a number of reasons. Villagers depended for water on the local wadi and one borehole well; but the wadi had been emptied by drought and the borehole suffered from mechanical failures. As the village is a stopping point on a livestock route, its resources of water, food and retail goods are under extra pressure. The villagers were also particularly concerned about damage to houses and fences by wind-blown sand.

An ERC mission to Um Inderaba in 1983 proposed that the villagers themselves decide what sort of forestry project they would undertake if they were provided with a SREP grant of 10,000 Sudanese pounds (£S). This caused initial confusion among the villagers, first because it was unusual for foresters to ask villagers what they wished to do and second because the villagers expected a more capital-intensive type of project. However, the village responded by submitting two proposals: one for the £S10,000 funding and the other for £S25,000, which included the rehabilitation of one of their two borewells. The smaller project was approved (although funds for the borewell rehabilitation were added later once the villagers' commitment had been demonstrated and initial goals achieved). During the implementation of the project the ERC foresters visited the village every two to three months for discussions and to assess progress, but otherwise the villagers were left to their own devices.

Environmental conditions at Um Inderaba were harsh: during the 1982-5 drought, for example, there was no rain at all in the area. Outmigration from the village was considerable and during this period the number of families fell from 2,000 to about 600. Despite these formidable obstacles, the following objectives were successfully achieved:

- a nursery was established and two seedling stocks raised
- a 3-feddan (approximately 1 hectare) windbreak was planted and protected, with almost 100 per cent survival of *Prosopis* species
- one feddan in the wadi was fenced off to examine natural forest regeneration in the absence of animal browsing
- a village guard rota was established to protect plantings (including a compensation scheme for the guards)
- *neem* trees were planted for shade, and *Acacia* and *Zizyphus* species for fuel and fodder.

The real commitment of the villagers to the project is demonstrable in the efforts that were taken to protect plantings. Despite the lack of rain and the large animal herds, each tree seedling was kept alive by hand watering with water transported by donkey cart. People built brick shelters to keep browsing animals away from plantings of *neem* (*Azadirachta indica*) around their homes for shade. Owing to the limited supplies of seedlings from the nursery, a penalty system was introduced by the village committee for any shade trees that died. All decisions were in the hands of the village committee because it held all the funds, although ERC foresters and (to some extent) Forest Administration foresters were available for advice and consultation. The fact that the villagers ran the project also meant that they could shop around to procure materials cheaply, so enabling them to complete below budget in a way which few external agencies could have done.

The plantings themselves did not in fact make a substantial contribution to welfare in Um Inderaba village. The windbreak, for example, is too small fully to control sand encroachment even when it matures. However, the most important feature of the project is the establishment of a new and innovative working relationship between forestry and village authorities and the establishment of a structure in the village which can work for other technical improvements to suit local needs and the local environment.

Other Community Forestry Projects in the Region

The Um Inderaba project can be contrasted with two other village forestry projects which were also supported by the ERC and SREP in the northern Kordofan region, at Um Tureibat and El Khwei. Despite far more favourable economic and ecological conditions, neither project was as successful, principally because no attempts were made to establish new institutional innovations as at Um Inderaba.

In Um Tureibat and El Khwei, foresters were permanently stationed on-site in the hope that this would accelerate project progress and facilitate its replication in nearby villages. A Peace Corps volunteer was posted to Um Tureibat, and a CARE forester to El Khwei. However, although nurseries were established, there was very little planting or community involvement in either village. Indeed, the continual presence of outsiders appeared to inhibit rather than facilitate local participation.

Source: Gamser (1987, 1988).[26]

12: THE MAJJIA VALLEY WINDBREAKS, NIGER

CARE and the Forest Service of Niger have since 1974 been engaged in an agroforestry project in the Majjia Valley of south-central Niger as part of the wider Tahoua Agroforestry Project. It was designed to protect and conserve soil, protect agricultural production and produce wood for use as fuel and poles. USAID has supported the project since 1981.

The Majjia Valley is home to an estimated 33,000 people, living in twenty-seven villages, and lies within the Sudano-Sahelian zone, an area of shrub savanna with a mean annual rainfall of 500 mm. The soils are comparatively rich and, as a result, population density is high at fifty-two persons per square kilometre. With the extension of agriculture (millet, sorghum, cotton and groundnuts) throughout the valley, much of the tree cover had been lost. Years of continuous cultivation leading

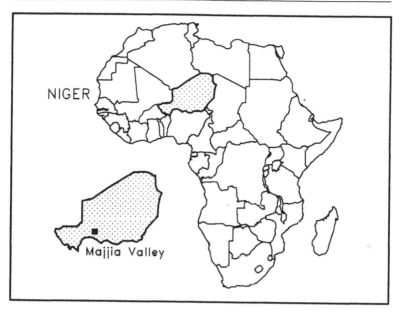

to declining soil fertility and erratic rainfall had resulted in marked declines in crop productivity. The rate of soil erosion from both water and wind had been increasing. During the long dry season from November to May, when little vegetation covers the ground, the desiccating harmattan wind blows almost incessantly down from the Sahara, carrying away valuable topsoil.

The project began when local villagers approached the government forestry officer (Daouda Adamou) in the town of Bouze for assistance in reducing the effects of wind erosion. The inspiration came from a few lines of windbreak planted a decade earlier further down the valley. In 1975 the first double row windbreaks of neem (*Azadirachta indica*) were planted. Since then more than half of the 6,000 hectares in the project area have been protected by almost 500 kilometres of windbreaks established in parallel rows 100 metres apart across the breadth of the valley.

All trees, in Nigerian law, belong to the government, and foresters have been cast in the role of "resource cops" to prevent farmers even from pruning the trees standing in their fields. Daouda Adamou pioneered a new approach to forestry for the Majjia Valley project, quite different from the old paramilitary style of the Niger Forest Service. He had built up excellent rapport with local villagers over

several years and had already persuaded many of them to plant
individual woodlots and trees. It was ensured that in the Majjia Valley
scheme farmers were treated as potential partners and the principal
beneficiaries of forestry work, rather than as potential criminals. The
villagers' confidence in Daouda, as well as the backing he received from
local authorities, contributed enormously to the success of the project
and the degree of voluntary participation in it.

Most of the labour for planting the trees, for instance, has been
provided voluntarily by the villagers, although initially some workers
were paid in food through a food-for-work programme. However,
foresters and Peace Corps volunteers choose the location of the
windbreaks and decide on tree placing and spacing, while paid nursery
staff raise the seedlings. The protection of seedlings is the farmers'
responsibility in the rainy season while they work in their fields, but in
the dry season paid guards take over the job. This guarding system has
common agreement by the villagers and complements their own
community sanctions on use of the trees. Hopefully, the paid guards
may in time become unnecessary. Villagers have also gained a good
understanding of the functions of the windbreaks and can show where
more are most needed. Indeed, the windbreaks are now spreading
spontaneously.

Part of this success can be attributed to the clear and direct benefits
from the windbreaks for the majority of local people. As early as 1980
studies in the valley documented positive impacts on cereal production
in the fields protected by windbreaks. Subsequently, a major technical
evaluation corroborated these positive results: see Table 1.3 (p.46). In
1984, for instance, a year of intense drought in the area, crop yields in
the protected fields were 18 per cent higher than in adjacent unprotected
fields, while in 1980 and 1985 the comparable increases were 23 per cent
and 20 per cent respectively. In a recent survey up to 72 per cent of
farmers said that the windbreaks increased food production. They have
also largely solved the wind erosion problem, which was their aim, and
have led to a significant growth in total agricultural biomass, including
husks and stalks for fodder, and a general increase in soil organic matter
to help improve its structure and resistance to erosion.

In addition, there have been significant gains in fuel and other wood
resources. An inventory of the oldest windbreak lines conducted in 1984
found that on average one kilometre of windbreak line contains about
110 cubic metres of wood. Based on current consumption estimates for
the Sahel, this amount could meet the fuel needs of about 250 people for

a year. Extrapolating further, the total amount of wood contained in the 1975 lines (10 kilometres) could provide enough fuelwood for about a tenth of the valley population for a year. These estimates, however, are based on clear felling rather than taking wood by loppings and toppings, which is how much of this need would actually be met.

However, not everyone has gained. To protect the young trees, large fines are imposed if any animals are caught in the fields. Rather than pay them, many livestock owners have sold their animals or moved them away. Herders have therefore been the main losers from the project. The other losers have been women, who have suffered the loss of grazing rights for sheep and goats.

Furthermore, many farmers were initially reluctant to give up some of their land to the project, although the villagers as a whole wanted to establish the windbreaks. In order to be effective, however, the project needed to cover a lot of land and required co-ordination between adjacent landowners. No compensation was made to farmers who may have lost a considerable area. Partly due to this, the sociological survey which formed part of the recent project evaluation found that 88 per cent of respondents thought that the trees belonged to the foresters and only 2 per cent thought the farmers themselves owned the trees.

On the whole, however, and particularly among local farmers, enthusiasm for the project has flourished, so that CARE has recently established a third nursery to help handle the demand for seedlings. The sum total of benefits is seen to outweigh the costs by far, although it has taken some time to reach this conclusion. During the first six years of the project, managerial difficulties and the economic hardships for the displaced herders were real problems, but after ten years the threshold of success seemed to have been crossed.

Moreover, the project is still being developed. Experimental felling in the older windbreaks is attempting to determine how best to manage them without seriously curtailing their protection function. Pollarding (removal of the tree crown alone) appears to be preferable to coppicing (removal of the entire above-ground portion of the tree) since it achieves a higher rate of regeneration and does not require the protection from grazing animals that was necessary with the coppice cuts. Regeneration from pollarded trees has been very encouraging, with substantial potential for producing wood on a sustainable basis.

The government has also agreed to a distribution scheme for this wood. It is shared equally between the landowners, those who participated in the cutting, and the village councils. This recent development seems likely to be successful, as long as local people, not officials, are the principal decision makers on how wood products are distributed and those groups who currently lose out also receive their share of the benefits. The resolution of uncertainties over tenure and cutting rights is perhaps the most important issue now facing the Majjia Valley project. It could be the key to both the sustainability and the wider replication of the farmers' initiatives.

Sources: Bognetteau-Verlinden (1980),[27] Delehauty, Hoskins and Thomson (1985),[28] Dennison (1985),[29] Williams (1985),[30] Gregersen *et al.* (1986),[31] Harrison (1987),[32] Jensen (1987),[33] Shaikh *et al.* (1988).[7]

13: THE HADO PROJECT IN KONDOA, TANZANIA

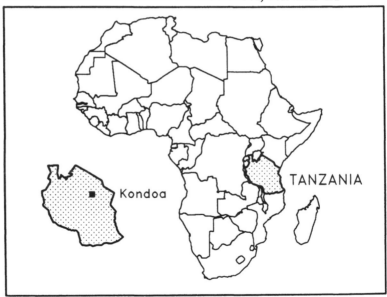

The Kondoa district in central Tanzania is a semi-arid area with annual rainfall below 800 mm and sandy loam soils which are prone to erosion when deprived of vegetation cover. One of the worst affected places is the so-called Kondoa Eroded Area (KEA), covering 12 per cent of the district, where the HADO project interventions have been focused.

The people of Kondoa depend on a mixed agropastoral economy, which is being threatened by over-stocking and over-grazing. The human population rose by 88 per cent between 1948 and 1978 and the animal population has increased alongside. The HADO sector has a far higher population density than the average for the district at 59 persons per square kilometre, according to the 1978 census, with a stocking density of 69 cattle per square kilometre. The condition of the soils and vegetation was disastrous and worsening, especially as more pastures were converted to arable production, so intensifying over-grazing on the remaining pastures.

The HADO Project

In 1973–4 a conservation unit named HADO (*Hifadhi Ardhi Dodoma* – conserve soil in Dodoma) was established with assistance from the Swedish aid agency SIDA under the Forest Division of the Ministry of Lands, Natural Resources and Tourism. Its objectives were similar to those of efforts in the 1950s to prevent land degradation, but with a sharper focus on rehabilitation of the "badlands" and a far greater emphasis on tree planting. The strategies were:

- terracing or contour ridging with tree planting for erosion control and eventual fuelwood provision
- establishing demonstration tree cultivation plots
- providing advice on soil conservation
- establishing tree nurseries, to include fruit trees for village plantings
- planting grass and shrubs for gully reclamation.

Hillside Closure

Some 11,300 hectares of degraded land in the KEA were treated but these efforts met with only modest success owing to the excessive density of cattle stocking. In 1979 the decision was taken to destock the KEA completely, and some 85,000 animals were transferred to lower-lying areas. This decision was entirely out of the hands of local people; the Forest Division was not prepared even to discuss any alternative suggestions. Indeed, the banishment was only achieved because it had the support of influential politicians who were willing to risk publicly supporting it.

Cattle owners constitute one-quarter of the local population, although many more households benefit from cattle than the number who formally own them.[34] Most of these people were strongly opposed to the destocking policy in 1979 and some were still opposed to it in

1986. No arrangements were made as to where the banished cattle were to go; in practice most were taken to relatives living in the plains to the south and east. Even today patrolling continues to prevent animals from entering the KEA, and a forester was once killed in a raid.

Benefits

By 1986 cattle owners were reconciled to the changed life-style and had recognized clear benefits from the improved farming conditions and the quite impressive regeneration of natural vegetation. In only two years just over half of the destocked area had regained ground cover and local perceptions are that "there is always some dead wood available for fuel collection these days". Moist depressions called *mbugas* which used to be covered with sand flows at the onset of the rains have been divided up for cultivation where two crops a year can now be harvested. They have been divided equally between people regardless of whether or not they had been cattle owners, leading to a progressive redistribution of wealth. Cash incomes are higher and nutrition has improved as people can now afford to consume dried fish from Lake Tanganyika. Although the area was previously a regular recipient of famine relief, this has not been necessary over the last few years.[35] Interestingly, there is no clear distinction between herders and farmers in their attitudes towards destocking or HADO.

In the last few years HADO's style in the whole Dodoma region has changed considerably towards a slower, "barefoot extensionist" approach, involving discussions with farmers and village meetings. It has even succeeded in getting one division, Mvumi, to destock on a voluntary basis. Its method was to go from village to village trying to get agreement, and to take village leaders to Kondoa to see what had happened there.

Tree Planting Efforts

HADO is also involved with forest plantations. This too has been controversial. Some plantations were inappropriately sited, for example on formerly cultivated and relatively good land.[36] This was not consistent with local land use priorities and has led to resentment, especially at Gubali Village, where plantations over a 300-hectare area have been regularly burned by the local people.[34] Communal woodlots proved to be a failure and have now been dropped due to their unpopularity.

The relationship between local people and the HADO foresters is at best a tense one. This is not surprising in view of the earlier history of Tanzania's Village Afforestation programme which was launched in 1967–8.[37,38] Forest reserves were created by gazetting land and evacuating the local inhabitants. Once established, the reserves were policed by government foresters and local people were denied access to what were formerly common-property or open-access resources. Against this background, assigning foresters to undertake the HADO project was bound to inhibit local participation in the planning process.

Yet despite these initial resentments, in some areas HADO's direct forest plantation activities are now gaining in popularity. Some 2,614 hectares of demonstration woodlots have been planted, from which people can purchase harvested timber and construction poles at competitive prices, while the remaining pieces are distributed freely as fuelwood.[34] The recent changes in HADO's approach have also led to work on an experimental basis with farmer tree planting, grassing of mini-dams ("bunds") and composting, among other things.

These changes by "official" foresters towards co-operative rather than restrictive attitudes could have encouraging results given the very great local interest in tree planting. A survey of farmers in the severely eroded areas under HADO jurisdiction found that 85 per cent of respondents were engaged in tree planting, while in moderately eroded areas the proportion rose to 93 per cent.[36] Moreover, there were clear indications that interest in managing those trees more effectively could be stimulated if multi-purpose tree species – which HADO or other agencies could help to provide – were used.

Sources: Östberg (1986),[34] McCall-Skutsch and McCall (1986),[35] Nshubemuki and Mugasha (1985),[36] Mgeni (1985),[37] Skutsch (1985).[38]

14: WOODLOTS OR FUELSTICKS IN KENYA?

In the early 1980s the Kenya Fuelwood Cycle Study (see Rural Case 9) tried to resolve a policy argument about the best approach to rural afforestation by carrying out an economic analysis of two alternative strategies. One was a conventional woodlot scheme favoured by many planners, the other a "fuelstick" approach. The latter was to grow shrubby tree species by direct sowing with harvesting once a year to

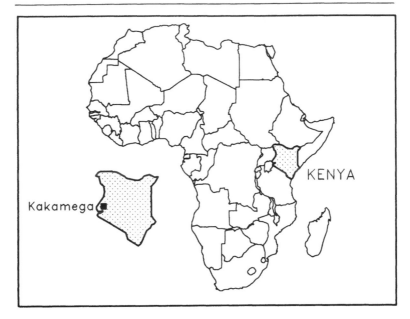

produce small "fuelsticks" which do not need to be split. The two model schemes were:

Fuelstick scheme: *Calliandra calothyrsus* planted on a one-metre spacing over an area of one hectare. Harvesting is by machete in Year 2 and each subsequent year until the shrubs are uprooted in Year 5 and replanted in Year 6.
Woodlot scheme: *Eucalyptus saligna* planted on a 2.5-metre spacing, thinned in Years 5 and 7, and clear-felled in Year 10.

Production and cash flow estimates for each scheme showed that if smallholders do not discount the value of money – that is, are indifferent as to whether they have one dollar today or in some future time – the conventional woodlot is the more attractive of the two options. However, smallholders generally have a very high discount rate, meaning that they would much rather have a dollar today than in a year's time or later.

    Richard Hosier therefore compared the two schemes at various discount rates, as shown at the bottom of the Table 5.3. Since the woodlots do not produce a big return until the tenth year, while the fuelsticks produce a revenue in most years, it was perhaps not surprising that the fuelstick scheme has a greater Net Present Value at all discount

Table 5.3: Production and Cash Flow for the Two Tree Projects

year	fuelsticks		woodlots	
	production (m³)	cash flow (K. sh.)	production (m³)	cash flow (K. sh.)
1		-2.66		-3.67
2	30	3.43		-1.6
3	30	3.43		-1.6
4	30	3.43		-1.6
5	30	1.23	20	1.72
6	30	-2.16		-1.6
7	30	3.43	30	3.39
8	30	3.43		-1.6
9	30	3.43		-1.6
10		1.23	200	32.08
Total	240	18.22	250	23.92
Net Present Value at discount rate of:				
0%		18.22		23.92
5%		14.11		13.12
15%		9.04		2.33

Source: van Gelder et al. (1983).[39]

rates greater than 4 per cent. The majority of smallholders, on the other hand, certainly adopt a much higher discount rate than this. The exercise demonstrates that it is small wonder that attempts to get smallholders to plant eucalyptus woodlots had mostly failed.

The fuelstick approach became part of the technical package used by the Kenya Woodfuel Development Programme (see Rural Case 9), especially when field work in Kakamega district later revealed that farmers were already practising agroforestry systems resembling the hypothetical fuelstick project. Woodlots were less favoured in the households which were surveyed, partly due to land shortages.[40] However, in south-eastern parts of Kakamega district, where urban markets for wood products in Kisumu and Kakamega towns were highly developed, woodlots were a more attractive proposition for local farmers.

Source: Hosier (1987).[41]

15: AGROFORESTRY IN KORO VILLAGE, MALI

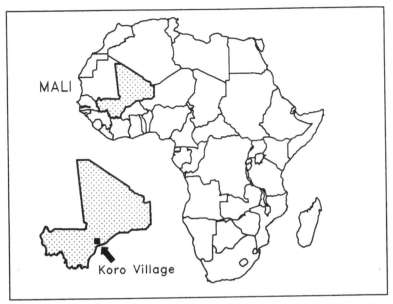

MALI

Koro Village

The Koro Village Agroforestry Project (VAP) in Mali's Fifth Region was implemented in mid 1986 by CARE with the support of the Norwegian Ministry for Development Co-Operation through its Sahel–Sudan–Ethiopia (SSE) Programme. The project has both technical and institutional objectives:

- to disseminate information about tree planting to increase farm productivity
- to identify locally appropriate agroforestry practices
- to develop and strengthen the extension capacity of the Malian Forest Service
- to establish the project as a pilot or model for other agencies, NGOs and the government of Mali.

The Koro VAP is engaged in extension for tree planting activities which include windbreaks, interspersed field trees, fruit and shade trees, and live fences. Nurseries follow the local, decentralized model, which is held to be a critical element in the long-term success of the project. Villagers already demonstrate a high degree of sophistication in tree culture. The CARE project staff took fourteen local farmers from Koro to the Majjia Valley in Niger to observe the success of their windbreak project (see Rural Case 12). The resulting adoption rate of tree planting among the Dogon people at Koro has been over 70 per cent.

Learning from the experience in the Majjia Valley, it was decided not to use paid guards to protect seedlings from livestock. Instead, extension agents were to decide on a village-by-village basis which local materials, such as shrub species, are most appropriate for tree protection purposes, taking the lead from the villagers themselves. The windbreaks, of which 18 kilometres were planted in the 1986 season and which had spread to twenty-one villages by June 1987, are considered to be community plantings even where they cross farmers' private land, owing to the village-wide scale of organization which is required for planting and tending windbreaks.

Of the other types of tree planting, fruit trees appear to be the most popular, but emphasis has now switched from planting them to the management of those already planted. Water is a limiting factor in this dry area, and while trees may survive they may fail to produce fruit unless cared for properly. Without good management, trees may compete with crops for space and moisture, since the suitable sites for tree planting also tend to be suited to dry season cropping. Competition with crops for light is also a factor to be watched: indeed, the crown size of an adult mango tree, typically 15 metres in diameter, may be larger than the cropped area in small plots.

Much was learnt about the management of live fences during the project. Careful management is required to create a true fence which does its job of keeping out animals, rather than being simply a "row of trees", but depending on species choice true fences may take some time to mature. The cost of establishing a living fence using seedlings from central nurseries may be around US$50, which is prohibitive for most villagers. However, by emphasizing cheaper methods of seedling production in decentralized mini-nurseries or direct seeding by farmers, the project has brought live fences within the economic grasp of most farmers – although they may still prefer to devote their efforts to planting and managing fruit trees.

Institutional Aspects

Extension efforts within the Koro VAP are very much based at village level. The regular staff meetings include a training component to emphasize ways of encouraging participation and local initiatives, and local relations have so far been very good. Following a full evaluation of the project in November–December 1986, which in itself did much to strengthen relations between CARE and the Forest Service, it was recommended that female extension staff be included to secure the greater participation of women in extension activities, which had to date been very low.

The project goals specifically include the institutional development of the Forest Service, which has not been a feature of other CARE projects in Mali. The more usual pattern has been to use the government's recently established local Development Committees as counterpart institutions. This has been very effective as DCs have good outreach and knowledge of local conditions, are able to reach decisions quickly, and already co-ordinate all technical service activities at the local level.

However, for the Koro VAP to work with the Forest Service is a new departure with several key advantages. One is that the project gains from being in close touch with policy issues, even to the extent that it can influence them, especially to support current changes in government structures, of which decentralization is one manifestation. The model will be particularly successful if it manages to go some way towards defusing the inherent tension in the CARE-forest service collaboration, which results from Mali's historical inheritance of a repressive, paramilitary forest police force.

In summary, the Koro VAP is providing valuable lessons about technical and institutional issues. CARE has incorporated into it the experience gained from many other projects and explicitly regards the project itself as a learning process. As yet the project is too new to evaluate properly, but since many of the fundamental ingredients of success for farm forestry projects are being integrated into the Koro VAP – putting people's priorities first, emphasizing sustainability through self-help, and recognizing the need for secure rights and gains – there are high hopes for its success.

Sources: Hagen *et al.* (1986),[42] Laumark (1987),[43] Shaikh *et al.* (1988).[7]

16: AGROFORESTRY DIAGNOSIS AND DESIGN IN KATHAMA, KENYA

ICRAF's agroforestry D & D methodology has been applied in the Kathama Agroforestry Project, in a community development setting in Machakos district of Kenya, for over five years. An important aspect of the two iterations of D & D as it was applied in Kathama is that they

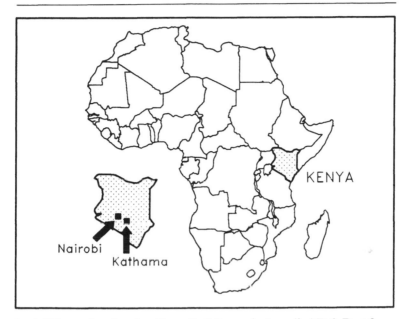

had different focal scales. Phase I of the project applied D & D at farm level[44] while Phase II was applied at watershed and community level.[45,46]

The D & D procedure need not always be applied in this way since the initial study could integrate both scales. However, in Kathama it was the farm-level D & D which suggested the need for a larger-than-farm-scale application of the method. In particular, the erosion problems experienced on many farms were caused at least in part by wider landscape processes and patterns, and the runoff from some individual farms affected other farms in the area.

It was also recognized that the individual household was not the only social unit capable of carrying out agroforestry trials, particularly in the light of a study of existing agricultural self-help community groups.[47] It was anticipated that community organization on the one hand and a broad approach to erosion control through landscape planning and design on the other[45] would be mutually reinforcing factors. However, the studies revealed that the social factors were in fact imperative for the successful implementation of agroforestry projects on any scale, whether at farm or at community level.

This first came to light over problems of water supply for raising seedlings on-farm. During the dry season this depended on public water sources, often quite distant from individual farms. Women were expected to carry the water but were reluctant to do so as they were not guaranteed any benefits from what were, in this case, men's tree plantings. The involvement of women as individual beneficiaries was obviously crucial to successful nursery management, while nursery work would also benefit if it was organized as a group activity near a permanent source of water on the farm of one group member. These groups were not truly "communal" but were inter-household working groups which provided a facility for reciprocal labour exchanges and the like.

The testing of possible agroforestry options suggested by the D & D exercise led to a change of emphasis within the project. Instead of planting hedgerows with crops for mulch (or "green manure" alley cropping) many farmers expressed a preference for hedgerow intercropping for fodder and fruit, and wider hedgerow spacing. In other words, they aimed to use tree biomass from hedges, and dispersed trees in grazing land, to improve the supply of green manure for their crops – a modification of their original "brown manure" strategy. Farmers with insufficient grazing land to support enough livestock to provide adequate manure were the most likely to prefer the new mulch and green manure strategy.

Sources: Vonk (1983),[44] Hoek (1983),[45] Rocheleau and Hoek (1984),[46] Wijngaarden (1983),[47] ICRAF (1987).[48].

17: THE VILLAGE AFFORESTATION PROGRAMME IN TANZANIA

Village forestry was proclaimed a national programme in Tanzania's Arusha Declaration in 1967. The programme took the form of seedling distribution by the Forest Division for farmers to plant in communal woodlots, and the planting by the division of demonstration woodlots in many areas with the help of paid labour. The main objective was to provide firewood.

A multi-media awareness campaign entitled "Forests Are Wealth" was launched in 1980–4, focused on eight of the driest regions in central and north-west Tanzania. News articles, posters and T-shirts, radio programmes, seminars and meetings were all used to spread the message to plant trees. According to the campaign report, the rate of planting rose from 4,500 hectares in 1979–80 to 8,500 hectares in

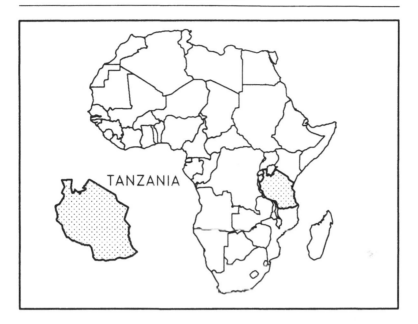

TANZANIA

1980–1. However, others have been sceptical of the precise role of the mass awareness campaign in this increased planting activity.[35] Tree planting seemed to be on the increase all over the country even before the campaign began. The increase registered in the campaign's first year may have been quite unrelated to the campaign itself, since very similar planting increases were observed in 1980–1 in both campaign and non-campaign regions.

The Village Afforestation Programme initially met with a mixed response from villagers. In the early 1980s, Margaret Skutsch interviewed people in eighteen villages to explore the reasons for the varying response to the programme.[49] Some of the villages had successfully established woodlots while others had not; some were in wet areas, some in dry; some were in wood-rich areas and some in wood-poor areas. Skutsch concluded that village forestry is being adopted in the villages of Tanzania, albeit slowly. Her findings are summarized here.

Why Some Villages Start Woodlots While Others Do Not

The concept of starting woodlots to provide a single main output – fuelwood – is an alien one in the livelihood systems of rural people with multiple needs and priorities. Where villages did not start woodlots it is very likely that any perceived need for extra wood could be satisfied by individual tree planting. Naturally the degree of need varies from one locality to another. There was an observed dislike among many villagers for the fuelwood-orientated species produced in the nurseries, since they clearly preferred multi-purpose tree species.

Starting a communal woodlot places extra demands on people's labour time, and may compete with the demands of agriculture. This will be resisted unless the need is proven and perceived, and cannot be satisfied in other ways. More importantly, the redesignation of private land as communal land which may be required for a woodlot is often strongly resisted by those who own or have usufruct rights to that land.

Another set of reasons relates to insecurity of rights generally. Lack of trust in village leaders, particularly in their financial morality, was often mentioned as a constraint to communal tree planting by villagers. Similarly, a mistrust of government motives was a significant factor. Some feared that the government sought to seize land, or redistribute settlements again, in the wake of the national villagization programme. In some cases, factionalism within communities inhibited communal involvement.

The attitudes of government forest extension officers were found to be very significant. Some were aware of the needs of villagers in terms of forest products and management, and were a primary cause of villagers' motivation to plant. In other villages this was not the case. The institutional structure of the Forest Service, and the close control of the party over the Village Afforestation Programme, have had a very profound and unintended impact on community tree planting activities.[50] Great stress has been placed on the establishment of woodlots as an end in itself, rather than as a means towards the community-level management of trees to meet locally felt needs.

Indeed, social forestry in Tanzania has become highly politicized. Villagers face the prospect of strong pressure to plant trees from both party officials and foresters, and consequently they may decide to plant a small but visible woodlot at the centre of the village. This satisfies not only the villagers' desire to relieve political pressure, but also the careerist motives of party officials and government foresters. The mere existence of woodlots in as many villages as possible – regardless of their real impact in meeting felt needs for wood products – is sufficient for foresters to secure the approval of their superiors or better chances of promotion.

Against this background, one of the most crucial findings of Skutsch's survey was that the reasons why villages failed to plant trees did *not* include a lack of technical knowledge, the inability to prevent browsing animals damaging seedlings, or even logistical problems in obtaining seedlings.

Why Some Woodlots Fail Once Started

Woodlots had failed in some 44 per cent of the villages in the sample where they had been started. Villagers in these places tend to lay blame on physical factors such as fire, pests or termites, and drought in particular. In part, woodlot failure can be attributed to the deliberate negligence of those who are supposed to benefit, owing to their being coerced into planting when there was no pressing need for a woodlot. The organizational ability of village governments was found to be strongly related to the degree of trust villagers had in them. Where both were low, villagers' awareness of the planned woodlot was minimal, and consequently participation was weak.

On the whole, village government control over the process of establishing a woodlot was good, and fines were imposed where cattle grazed seedlings, or wood was stolen for firewood or poles. A corollary of such control, however, is that the trees and tree products were considered to be the property of the village government, which again inhibits local participation. In villages with well-organized woodlots the local government was responsible for drawing up a distribution plan, and for the sale of poles and fuelwood back to the villagers. The revenues were then contributed to the village general fund. Not successful everywhere, this management system depends critically on the organizational ability of the village government, which tends to be low in villages with a markedly heterogeneous population.

Planting did not always continue once woodlots were established, partly because the lots were rarely able to meet their intended purpose. Most of the woodlots in the study were too small to have much impact on satifying fuel demand. A 2-hectare stand or *Cassia siamea*, for example, would provide sufficient fuel for only two to four families. Each family would need about half a hectare, or more in relatively high altitudes where space heating requirements are higher. Most of the village woodlots are only 2 hectares for an average of 450 families; the largest was a mere 12 hectares.

Here was identified a significant extension gap to be filled. The concept of rotational forest management did not seem to be familiar to village leaders, nor was there a clear understanding of the area of woodlot necessary to supply village needs. What is required in villages with woodlots is a system of incremental planting, together with a woody biomass harvesting plan based on rotations of varying lengths according to the desired products, whether they be poles, fodder, fuelwood or any other identified need (see Chapter 1, "New thinking on tree economics", pp.47–53).

In the light of these experiences, the strategy of the Village Forestry Unit within the Forest and Beekeeping Division has begun to change considerably. The present emphasis is on decentralizing nurseries, the use of various cost-saving techniques in nursery practice, and a shift towards woody biomass management rather than tree planting. The former includes, for example, encouraging the natural regeneration of existing vegetative cover and the direct sowing of seeds rather than conventional seedling propagation. The potential economic impact of the new strategy is also discussed in the section on the economics of rural tree growing in Chapter 1.

Sources: Skutsch (1983),[49] McCall-Skutsch and McCall (1986).[35]

18: AGROFORESTRY OUTREACH IN HAITI

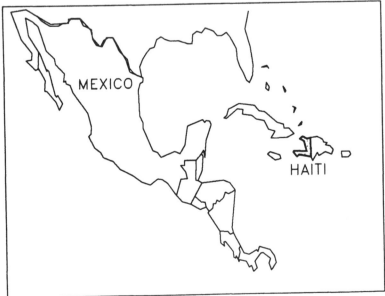

In 1981 USAID began to support the extension of agroforestry in Haiti by providing funds to three international private voluntary organizations (PVOs): Pan American Development Foundation (PADF), Operation Double Harvest (ODH) and CARE. The plan was to make a radical departure from the traditional approach to soil conservation through public works programmes by using trees both to protect the soil and increase farm incomes. The basic strategy was simple: to provide millions of cheaply produced tree seedlings to farmers and ensure that the farmers owned the trees and benefited from them.

The three PVOs developed different systems for engaging farmers in tree planting. PADF supplied seedlings across much of the country by working through the many small, local and international NGOs scattered throughout Haiti. They received the largest funding but used this to make sub-grants to the more than 170 NGOs who eventually worked with them. ODH worked with farmers who had leaseholds and, for the first few seasons, provided nursery seedlings for PADF and CARE efforts. Its work involved relatively large-scale tree-farm demonstrations and has been criticized for not always applying the most appropriate technology and for inadequate research.[51] CARE worked in one of its traditional programme areas, the north-west province, directly through its own extension system.

The PADF and CARE components provided the bulk of the 27 million trees which 110,000 farmers planted during the programme's first five years. Both agencies provided seedlings grown in small containers that greatly reduced the cost of seedling production and transport, even though containers and potting mixes were imported. Seedlings were distributed free, followed with extension advice on their planting and maintenance.

In particular, the extension agents took trouble to assure the farmers that the seedlings belonged to the planter, that they could harvest the trees when they chose, and that they would receive the benefits of the harvest. Farmers were at first offered an incentive of US$0.10 for each seedling which survived one year after planting. However, this was quickly dropped as the demand for seedlings far exceeded original expectations. But despite this quick take-off, USAID's own Audit Report of the programme drew attention to the lower-than-expected tree survival and the stunted growth of the seedlings, with survival rates said to be only 45 per cent, not 70 per cent as predicted.[51]

Farmers grow the trees principally as a cash crop, but for investment rather than income. They are planted as a form of savings, since cultivating trees is one of the few opportunities many rural Haitians have for accumulating assets.

Farmers have controlled the programme in two principal ways: species selection and choice of agroforestry configurations. For instance, a third of all planting in recent years has been of local hardwood species, rather than the exotic species initially grown by CARE and PADF. The farmers selected the agroforestry practices that appealed to them from a range of options included in the extension programme. For example, though not in the original plan, hillside hedgerows of direct-seeded *Leucaena* species are becoming very popular.

Given the success of the programme to date, USAID has recently extended funding for another three years. The rural landscape is visibly changing as new farm forestry develops. New income-earning opportunities have also developed; for example, farmers have begun to harvest the older trees for charcoal. The Agroforestry Outreach Programme has taught two main lessons: first, that donor organizations can distribute large funds by supporting "umbrella" NGOs to manage and fund other organizations working at the "grassroots" level; and second, that it can be crucial to make it clear that the people who work on tree planting and management will reap the benefits.

Sources: Conway (1987),[52] Kramer (1987),[53] USAID (1984).[51]

19: CO-OPERATIVE FOREST MANAGEMENT IN GUESSELBODI, NIGER

There are few examples of community forestry projects designed to enhance existing resources. Generally, projects clear natural "waste" and establish plantations at great cost, promising high returns. In many areas, such plantations have been costly failures. In Bandia, Senegal, for example, the cost of establishing a eucalyptus plantation was US$800 per hectare, although this area produced only 1.5 cubic metres of wood a year. Yet in the same region, the native scrub (*Acacia seyal*) produces an average of 1–1.5 cubic metres annually per hectare (range 0.8–3.2 cubic metres) with no management and, of course, no establishment costs.

In 1980, USAID and the Forest Service of Niger began an innovative project designed to manage 5,000 hectares of the Guesselbodi reserve

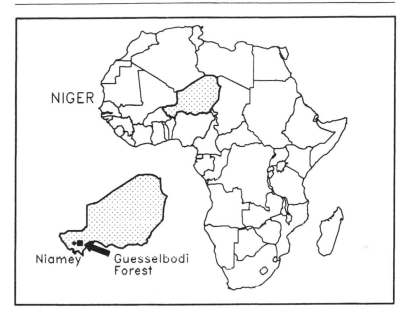

forest using what it termed "participatory" management, involving local residents. This was part of the Forestry and Land Use Planning (FLUP) project, which has many components including research, soil conservation, agroforestry, establishing a forest co-operative, and grazing management. Two years of preparatory field work based on discussions with residents as well as soil and vegetation mapping were conducted before a management plan was adopted. Information was collected on past and present uses of the forest, changes which had occurred over the past thirty years, local priorities and preferences for forest management, and local customs regarding forest use. The forest was found to be severely degraded (40–60 per cent of forest cover had been lost during the past thirty years) with severe soil erosion.

The management plan divides the area into ten parcels to be managed on a ten-year rotation, based on a coppice cycle of the dominant *Combretum* species. Each parcel will be protected from grazing on a three-year rotation, although residents are permitted to enter restricted areas for forest products such as gum, food, medicine and to cut hay.

Management is by a co-operative from the nine forest villages, which the project helped to establish. Each village elected five officers who together make up its operating committee. After two years of negotiations, a contract was signed with the government which accords the co-operative the right to exploit the forest as long as it abides by the policies set out in the management plan. The co-operative has responsibility for firewood cutting and the sale of grazing permits, while the Forest Service directs the ongoing conservation and restoration programme. All wood cut in the forest (at present by about seventy cutters) is obligatorily sold to the co-operative.

Part of the success of the co-operative and the project is due to the market opportunities offered by the capital city of Niamey, which is only 30 kilometres away on a good metalled road. Fodder and firewood fetch high prices there, and without this market incentive and the relatively high existing level of commercial development and skills, it is unlikely that the co-operative would have worked.

Preliminary analyses of the project's economics suggest that establishment costs are about US$320 per hectare and that firewood sales are already covering 80 per cent of the project's recurrent costs (mainly salaries for the forest guards).[54] Most of the cash generated by the sale of hay and grazing permits will therefore be profit for the co-operative. Over the first twenty years, the expected net returns in millions of CFA francs are: 54.8 for hay sales, 2.7 for hay-cutting permits, 46.6 for firewood sales, 2.8 for firewood permits, and 31.5 for grazing permits: a total of 138 million CFA francs (or US$310,000 in 1985 currency terms). The forest service shares in the profits, which go to its natural forest management fund, partly to compensate for the loss of lucrative fines previously imposed in the forest for illegal cutting.

Based on these results, similar management enterprises are planned along all the main roads leading to Niamey and other major urban centres in Niger. Eventually, it is hoped that all forest products sold in the major urban centres should come from forest enterprises that are managed and controlled by local populations.

The important features of this project are that local people were involved from the early planning stages, their interests were respected (for example, by allowing access to forest products during restricted grazing periods) and their participation sought for the management of the forest areas. Thus, through the co-operative local residents have been given responsibility for managing their local forest resource, while still receiving support (for restoration and conservation) from the Forest Service. At the start, researchers knew little about the ecology of the forest area, but they relied heavily on the knowledge of local foresters and people. The initial constraint itself – lack of knowledge about the ecological processes of the area needed for the management plan – actually fostered a two-way information flow between researchers, planners and local forest users.

Sources: Heermans (1987),[54] Falconer (1987),[55] Taylor and Soumare (1984, 1985),[56] Thomson (1981),[57] Shaikh *et al.* (1988).[7]

20: REFUGEE FORESTRY IN SOMALIA: THE "STEP" PLAN

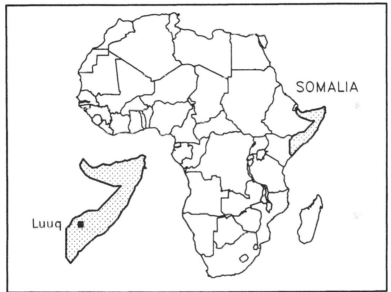

Refugee forestry projects pose particular problems because the target groups do not expect to be refugees for long or to be around to reap the benefits of any trees that they plant. To generate community involvement in forestry under these conditions might seem an impossible task, but a project in Somalia has done so by focusing on the provision of short-term benefits from tree planting activities.

In the late 1970s and early 1980s a war in the Ogaden Desert in the Horn of Africa, coupled with a severe drought, forced hundreds of thousands of people to leave their homes and seek refuge in camps established by the Government of Somalia. After the worst of the crisis was over, InterChurch Response for the Horn of Africa (ICR), a consortium of international church-based relief agencies, started longer-term development activities.

In 1982 a forestry project was initiated to try to generate community involvement within refugee camps in Luuq district of Somalia. ICR developed what they called the "Step" Plan to encourage the camp residents to plant trees, hoping to increase fuelwood supplies and do something to slow environmental deterioration. Within a year 2,000 refugees were participating, four times more than ICR expected, and over 90 per cent of them women. In fact, ICR could not accommodate everyone who wanted to plant trees.

The method they used was a progression of successively more difficult and rewarding tasks, which people follow step by step, at their own pace. There are eleven steps in the full plan, beginning with a simple introductory talk and the distribution of seeds by an extension agent who visits people in their homes. Gradually people learn more in the courses and receive more seeds and plastic seedling bags if they are successfully planting. Surviving trees were purchased by the project for 2 Somali shillings each.

After Step 4 there are three alternative tracks to follow: land preparation, nursery work, and tree planting. Payment is made for the work in the form of simple tools for land preparation and tree establishment (trowels, shovels, etc.). The cash payments for surviving trees and the tools provide the short-term benefits.

By starting off small, the project was self-selecting of those interested. Anyone could enter Step 1 but to continue they had to succeed at each step. Fewer than 15 per cent of people starting in Step 1 went on to Step 2, but the follow-on to further steps was over 80 per cent. The project could produce large benefits to participants, but they had to be involved for nine months to earn 240 Somali shillings (US$15). By this stage people had had full training and the project transferred some capital to them which could be used for other purposes as well as forestry.

The nursery was controlled by a committee composed of local government officials, ICR staff and Step Plan participants. They issued free permits for collecting animal forage from the woodlots, which would also provide wood products. Control of the woodlots is to be handed over to participants, but it remains to be seen how successful this will be. The Government of Somalia has supported the transfer of user rights to the people who planted the trees, since this is in line with the system of land tenure established for agriculture in Somalia.

The Step Plan is an interesting way of generating community participation slowly. Participants earn cash, while the other benefits in the form of training, and tools stay with them when they leave the area

– even if the trees stay behind. An indication of the project's popularity was the strong resistance of women and the poor who took part in it to later attempts to hire project labour from among men in the camps.

Source: Orr (1985).[58]

21: PUTTING TREES INTO NON-TREE PROJECTS IN KENYA

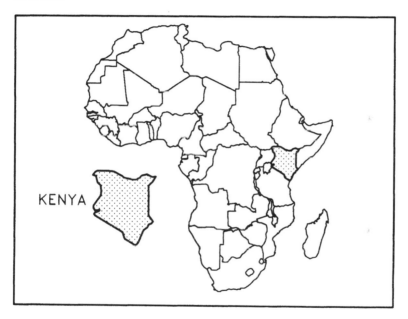

The potential for increasing woody biomass by indirect approaches can be illustrated by a few examples of Netherlands-assisted projects in Kenya where options have already been identified.[59] These include the Dairy Development Project, implemented by the Ministry of Agriculture and Livestock Development; the Rural Access Roads Programme; and the Arid and Semi-arid Land Development Programme.

1. The *Dairy Development Project* already involves small-scale experimentation with fodder trees in most of its project areas. There is a co-operative agreement with the Kenya Woodfuel Development Programme in Kisii district for advice and seedling supply. Its objective is to improve milk production in the high potential farming areas, where a limiting factor is inadequate protein in roughage feed for cattle.

Fodder trees could supply extra protein from leaves, pods and young twigs. Suitable species in terms of protein content and ecological conditions include *Leucaena leucocephela, Calliandra* species and *Sesbania* species.

2. The *Rural Access Roads Programme* aims to link areas of small-holder agriculture with markets to help generate income and stimulate rural development. One component of road construction is the building of small check-dams to facilitate drainage and prevent erosion, for which purpose trees are grown. Appropriate species could be selected for use in the programme and would also provide seed to passers-by for on-farm tree planting. Species choice would be governed by the particular requirments of local people.

3. Some 80 per cent of Kenya's land area is classified as arid and semi-arid and mainly supports pastoral groups. The *Arid and Semi-arid Land Development Programme* is focused on thirteen districts and receives external support from nine major donors, each with a restricted district focus. By contrast with farmers in the wetter, high potential areas the indigenous knowledge of pastoralists tends to concern the use rather than the propagation of trees. Techniques for tree culture cannot therefore simply be transferred across from more humid areas. The main potentials for trees within the programme are fodder production for drought-period feed; charcoal production for regular cash incomes; and windbreaks for crop protection since pastoral groups are increasingly engaged in sedentary agriculture.

Source: van Gelder (1987).[59]

22: THE NAAM MOVEMENT IN BURKINA FASO

The *Naam* movement in Burkina Faso is based on the traditional Mossi village organizations from which it takes its name. The *Naam* groups are self-help co-operatives, based on principles of equality, with leaders elected by the villagers who are familiar with them. The long tradition of self-help among these groups on the Yatenga plateau was an important factor in the successful spread of OXFAM's water harvesting project which uses rock bunds for soil and water conservation to increase crop yields (see Rural Case 10). The extension and development method of this project epitomizes the approach of the *Naam* groups themselves – learning by seeing and doing – and they are an

important channel for farmer-to-farmer training in the construction of rock bunds.

Naam groups address the full range of people's felt needs and are involved with a correspondingly wide range of development activities, from primary health care to wood resource management. Coming from self-help groups with their own rationale, such indigenous initiatives should be sustainable.

The *Naam* movement was founded in 1967 by Bernard Lédéa Ouedrago, a teacher turned rural development facilitator, in an effort to find an alternative rural development approach to top-down *animation* or consciousness-raising. The idea was for villagers themselves to define local problems and appropriate solutions, building on indigenous knowledge. In 1976, Ouedrago established an umbrella organization to co-ordinate the activities of the *Naam* groups, which had been steadily growing in number. By 1985 there were 1,350 *Naam* groups, in seven regional federations. The umbrella organization is the Six "S" Association (*Se Servir de la Saison Sèche en Savanne et au Sahel*), which provides technical and financial back-up for local *Naam* groups, and raises international funds for certain indispensable imports, such as medicines, hand tools and pump equipment. This type of assistance is imperative in a rapidly changing rural environment so as to empower local communities to manage the process of change and development in their own best interests.

Such grassroots development can have a surer momentum than the initiatives of outsiders. It can have significant impact, and similar organizations have been created in Senegal, Mauritania, Mali, Niger and Togo. The real issue is whether or not it is possible to galvanize these kinds of institutional mechanisms for popular mobilization in rural development in those areas where they are not spontaneously emerging. Where they are already strong, as in Burkina Faso, it is certainly possible to provide outside support to enhance the effectiveness of their initiatives.

Source: Harrison (1987).[32]

23: PROJECT CAMPFIRE, ZIMBABWE

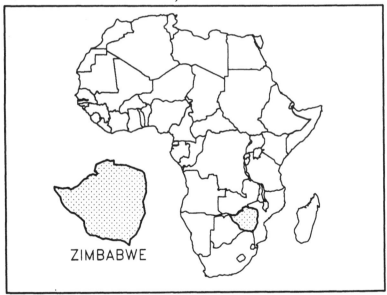

Project Campfire, in Zimbabwe, is a programme under the Department of Wildlife called the Communal Areas Management Programme for Indigenous Resources. It is an innovative approach by the Government of Zimbabwe to facilitate local action in natural resource management throughout Zimbabwe. It attempts to formalize the process of supporting indigenous development initiatives and embodies the important guiding principles that:

- local communities themselves are the custodians of their natural resources
- a flexible approach allows for learning from mistakes
- success depends on: (1) getting the technical package right; (2) on resolving the conflicts of interest that inevitably occur within communities; and (3) control by communities over their own destiny by developing viable systems for managing marginal areas.

The main elements of the programme are:

- communities join the programme voluntarily after discussion with extension workers
- communities themselves define which assets are individually owned and which are common property resources; clear definition of communal areas is vital
- a community co-operative is set up for management of the shared assets, and a management plan drawn up
- the programme offers certain incentives at the outset; e.g. electrified game fencing to protect arable holdings
- co-operatives have their own administrative boards with all adults as shareholders
- a pyramidal institutional structure is developed: all community co-operatives unite under a regional board, with representatives of technical agencies and district officials; regional boards combine under a national board, co-ordinated by the Ministry of Natural Resources
- communities are given asistance in organizational, technical and financial matters.

The approach of Project Campfire has yet to be properly tested in the field but does provide an interesting model that might also have a good potential in other countries.

Source: ETC (1987).[1]

24: RAPID RURAL APPRAISAL IN WELO, ETHIOPIA

A recent application of Agro-ecosystem Analysis was carried out in 1988 by the Ethiopian Red Cross Society in the Welo region of Ethiopia. The RRA team analysed two Peasant Associations (PAs), Abicho and Gobeya, in terms of their levels of diversification, a factor

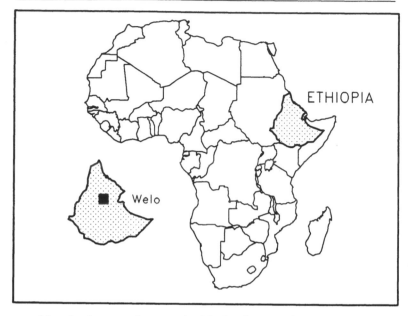

considered to be central to sustainable development in the project areas, as it makes for greater livelihood security, increased productivity and better levels of nutrition and health, thus acting as a development catalyst for the PA.

The RRA took a total of eight days, alternating between workshop sessions and visits to the PAs. For each PA, diversification was examined both in space and in time. The RRA team split up into subgroups, each with different tasks.

The techniques used to examine diversification in space included the drawing of maps and transects. In particular the team focused on some of the home gardens in each PA. Diversification in time was examined by means of calendars to investigate the seasonality of rainfall, crops, livestock management and breeding, pests and diseases of crops and livestock, prices of the major agricultural outputs, labour requirements (for both men and women), types of food eaten, human diseases and social events.

For the Gobeya PA, the analysis of the complete seasonal calendar revealed that there was a critical period of the year when a number of problems coincided. During June, July and August, when labour demands are heaviest for planting and weeding the main season's crop, crop pests are a particular problem, the shortage of fodder most acute, diarrhoea debilitates livestock, the incidence of human diseases reaches its peak, and at the same time the purchasing price of sorghum to local farmers is high.

The different perspectives of a range of disciplines were brought to bear on the problems of Gobeya PA during this process. They included health care, agronomy, plant pathology, sociology, anthropology and livestock management. This eclectic approach was essential in showing how individual problems combined to create a critical bottleneck at a particular time of year.

From the lists of specific problems and opportunities, the Gobeya RRA team chose four key issues which seemed to warrant further investigation. These were water utilization and management; crop rotation; the use of food and water (health and marketing); and the development and utilization of forests. These issues were examined in greater depth by means of interviews with groups and with different types of individuals, new diagrams, and ranking exercises.

For example, discussions around tree-related problems led to the identification of the key issue of forest development and utilization. The specific problems included the attraction by farm trees of birds which attack cereal crops; the eating by grazing livestock of trees planted by the lake shore; and the low survival rate of seedlings because of: (1) shallow soil; (2) inadequate spacing between the seedlings; and (3) the inaccessibility of the sites, discouraging proper cultivation and maintenance.

The team employed a simple ranking procedure to find out which tree species were preferred and why. Farmers were asked to compare, in pairs, six of the most widely used reforestation species, until all possible combinations were exhausted. The species were *Eucalyptus camaldulensis, E. globulus*, African olive (*Olea africana*), juniper, white acacia, and *Croton* species. The farmers were asked which species was "better" in terms of usefulness, and why. Ranking techniques are helpful in showing up the differences in perspective of different members of a community, as the team found in Gobeya.

A forestry extension specialist was then asked to make pairwise comparisons of the same species, with particular regard to their characteristics in terms of ease of nursery cultivation, establishment, productivity and erosion control. The forester's ranking and the farmers' ranking were compared. Whereas the forester ranked the two eucalyptus species first and second, followed by the African olive, the farmers ranked African olive first, followed by the eucalyptus species.

Ranking techniques are invaluable in highlighting the different perspectives of insiders and outsiders in local community development, as well as of different people within the community. The rankings of the extension worker and the farmers provided different but equally important and complementary perspectives on the choice of reforestation species. One conclusion drawn from this was that a tree which combines the fast-growing characteristics of the eucalyptus with the versatility of use and better erosion control of the African olive would be of great value.

The focus here on the key issue of forest development and utilization is not meant to imply that this was the most important priority identified by the Agro-ecosystem Analysis. Seven "best bets" were identified, of which two dealt directly with this issue. The shortlist of "best bets" to address people's expressed needs in Gobeya, not necessarily in priority order, was:

- drought-resistant, early-maturing crops
- home garden development
- small-scale irrigation from springs
- reforestation for soil and water conservation
- agroforestry for multiple purposes
- credit system
- strong primary health care committees.

Finally, the RRA team jointly assessed the likely impact of each best bet on the productivity, stability, sustainability, and equitability of Gobeya livelihood systems, and considered their relative cost, time-scale and feasibility. To the extent that this did distinguish between the best bets on these criteria, home garden development appeared to be the most promising option for Gobeya PA. But the importance of site-specific analysis cannot be played down. The most promising best bets for neighbouring PAs might be completely different.

Source: Ethiopian Red Cross Society (1988).[60]

Part II

Urban Centres

Chapter 6

Paying the Price

When we turn from the countryside to the cities, woodfuel problems and the means of solving them change quite radically in scale and kind, as do the contexts in which these issues are set. This is particularly the case for the larger cities which, in Africa, account for most of the urban population and the most pressing energy problems.

Most obviously, urban consumers pay for their fuels and do not produce them. As with food and other biomass commodities, cities are concentrated centres of demand which draw their needs from dispersed rural sources. The complexities of the rural production-consumption systems which we examined in Part I are therefore replaced by consumers, producers and the all-important commercial distribution systems which link them. In some ways this simplifies the picture, since issues of consumption and production can to a large extent be addressed separately and quite directly with more focused goals. Furthermore, the actors tend to follow the rules of economic behaviour in which supply and demand are driven by familiar factors of income and prices.

However, the picture is also complicated by the buyers, transporters and sellers of the woodfuel market. As we shall see in this chapter, the market and its traders are the pivots on which depend almost every option for providing adequate and sustainable woodfuel supplies while safeguarding the consumer from intolerable price increases.

The second obvious difference is the political and economic dominance of the larger cities which means that – at least for the time being – they can command the resources they need, reach for them over vast distances, and pay little or nothing for their production or replacement – thereby imposing substantial costs on rural people and the natural resource base. Rapid population growth, itself a reflection of urban dominance, can only intensify these pressures.

The basic context of urban woodfuels is therefore a three-cornered conflict between the needs of urban consumers, rural producers and the natural resource base – with the trading system which links them all holding the centre of the ring.

On the one hand, the mass of poor urban consumers cannot tolerate substantial (or any) increases in woodfuel prices. But their needs are growing and the traders who are in business to meet them will always try to operate with the lowest costs. Consequently, there are mounting and almost irresistible pressures for them to take wood from any unguarded or "open access" sources they can find or, if they must pay, to get it on the cheap. In the worst cases, where trees are scarce, this can lead to severe conflicts between the traders and rural people. More generally, it tends to put a price on what were previously "free" resources, usually to the detriment of the rural poor in particular.

On the other hand, the urban market and the commercialization of wood opens up many opportunities for rural people. Farmers can earn cash by producing wood for sale to urban centres as firewood or for charcoal making (or construction poles) but only if they are within economic reach of the markets. As for the forests, most of the vast numbers of trees which are felled to clear new farm land or in "slash and burn" agriculture – resources which typically greatly exceed local rural needs – never reach the city due to market factors. Labour shortages on the farm combined with low prices for the trees and remoteness from the city often ensure that most of the trees are burned rather than sold to the urban centres.

Although there are many possible and very positive solutions to these conflicts there have been few successes while several promising approaches remain untried. As with the rural areas, this general failure to solve the key urban woodfuel problems has to a considerable extent been due to naive diagnoses of the basic issues and their broader contexts.

We therefore begin this chapter with a sketch of the underlying contexts of urban woodfuels. We then draw the threads together by presenting a summary of the main objectives and options for resolving urban wood energy problems, using it to signpost the contents of the remaining chapters. Having set the scene, most of the chapter then looks at the central topics which link most of these issues together: woodfuel prices and the trading systems which govern them and the link between the cities and rural resources.

URBAN ISSUES AND CONTEXTS

Growing Cities

Most towns and cities in Africa are growing very rapidly. For instance, in the nine countries of the SADCC region, from 1975 to 1980 urban populations grew at a rate between 6 per cent and nearly 13 per cent a year, or about twice to six times faster than rural populations.[1] But it is the largest cities which are growing the fastest. In sub-Saharan Africa the population of cities with over a million inhabitants nearly trebled between 1960 and 1980, from 16 million to 46 million, while the numbers living in cities of 100,000 or more rather less than doubled from 58 million to 96 million.[2] Perhaps more strikingly, in 1980 the largest city accounted for one-third to two thirds of the entire urban population in eighteen out of twenty-three sub-Saharan countries with adequate census data, while in Guinea and Mozambique the largest city had 80 per cent or more of the total urban population.[3]

At the simplest level, this means that urban woodfuel use is also rising very rapidly and becoming highly concentrated. Much of the growth in the largest cities comes from poor people migrating from rural areas, bringing traditions of using woodfuels and, like the rest of the urban poor, unable to afford alternatives. While woodfuel consumption is typically lower than in rural areas, this is offset in many cities by the widespread use of charcoal which, with present technologies, requires roughly twice as much tree wood to provide a unit of energy as does firewood.[4] Urban woodfuel demands on tree resources are therefore probably growing at least as fast as urban populations and in some countries could well exceed those of rural areas within the next twenty years.[5]

Effective and far-seeing woodfuel actions are therefore urgent for most cities, either to prevent today's problems from worsening or to avoid the emergence of serious problems in places which are now relatively free of difficulties. The concentration of demand into the big cities also tends to intensify the pressures on tree resources in their hinterlands and along major transport routes, and amplifies alarmingly the scale of the actions which are needed to provide adequate and sustainable supplies.

Dynamic Cities

A second basic urban context is the economic and social dynamism of cities and towns. The larger cities in particular are the gateways to the global economy, the channels for modern influences, goods and services, and the centres of investment, employment opportunities and cultural change.

These characteristics are reflected in the wide choice and flexible patterns of urban fuel consumption. The most striking and persistent feature for the household sector, which dominates urban woodfuel use, is a transition with increasing income from firewood to "modern" fuels, with their greater convenience, cleanliness, versatility and inherent efficiency. The informal commercial and artisanal sectors – cafés, bakers, brewers, metal workers, potters and so on – also tend to switch from woodfuels to modern fuels as their prosperity rises.

We look more fully at this energy transition in Chapter 8. At this stage, the essential point is that underlying this broad trend is a much more volatile pattern of fuel use which can change rapidly, almost on a day-to-day basis. In nearly every city, modern fuel supplies are erratic or sold in only a few places, especially in the poorer parts of town. Woodfuel prices can vary as much as threefold in different parts of the city at the same time.[6] Access to the fuels which are available or relatively cheap depends on the time that can be spared and the available transport for taking them home – whether on foot, bicycle, donkey cart or a friend's truck. If cash is short, firewood (but not charcoal) can often be scavenged for "free" by the urban poor from timber yards or discarded packaging in the city, from trees in the urban hinterland, or from rural friends and family on weekend bus trips, and so forth. But again, these opportunities depend on time, transport and other highly variable factors.

Short-term fuel switching and changes in patterns of buying or gathering fuel are therefore common. To minimize the risk of not having fuel for essential tasks such as cooking, urban people put a high premium on security of supply and frequently have the equipment to use two, three or even more fuels. They can also choose not to use fuel at all by eating out or buying from the hot-food stalls that crowd many urban streets. The informal traders of the woodfuel market can also adapt quite rapidly to changing circumstances by altering the means, costs and profits of woodfuel transport and distribution.

In short, urban fuel use is complex as well as flexible and dynamic as prices and supply conditions alter in the city as a whole – or even in different parts of it. Rapid changes in the price and availability of other goods, and of job opportunities and income, add to this dynamism.

Two major implications for urban woodfuel initiatives follow from this. First, within the cities fuel problems and opportunities for change are often highly localized in nature and – as in rural areas – specific to people and places. Actions which seek to address the needs of the urban poor must be based on a good understanding of how the poor live and must often be targeted on specific areas or particularly vulnerable social groups. As in rural areas, this puts a high premium on consulting people and community groups in order to place their energy problems in a total context as well as picking up and helping to reinforce positive changes that have already begun. The urban poor, no less than rural people, often have an extraordinary ability to adapt, innovate and initiate communal self-help measures.

The second implication is that huge schemes are often based on data about urban energy patterns which are hopelessly generalized and out of date. In many cities, the only baseline data for such schemes might be a single survey (or worse, several conflicting surveys) on energy consumption patterns, fuel prices and so forth, made several years previously on a small sample in one week, with the results then averaged for the city as a whole. Such information obviously misses out entirely on the dynamics of urban energy use over time but also ignores the finer but often crucial spatial and class variations.

Falling Incomes

One crucial aspect of this dynamism is that urban economies and the labour market have undergone massive changes during the last decade. In most sub-Saharan countries real wages have declined drastically, regular wage employment has become more precarious, and the distinction between formal and informal employment has become blurred. Second jobs or "moonlighting", barter and informal "dealing" have become a basic way of life for many urban people while even the professional classes, such as senior academics and civil servants, often earn so little that they must leave work early to grow food or raise a cash crop on their suburban or peri-urban homestead "farm".

The drastic decline in real wages is illustrated in Table 6.1, using information from the International Labour Office.[7] In Ghana and Sierra Leone real earnings fell by around 80 per cent between 1975 and

Table 6.1: Decline of Real Non-agricultural Earnings in Sub-Saharan Africa

Real earnings of employees in most recent year compared with earlier year:

Sierra Leone	1972–1985	18%		
Ghana	1975–1982	22%		
Tanzania	1972–1983	36%		
Zambia	1972–1985	57%		
Malawi	1972–1985	63%	(1982–1985	68%)
Swaziland	1972–1985	63%		
Kenya	1972–1985	68%		
Zimbabwe	1972–1984	111%	(1982–1984	80%)

Source: Van Ginneken and International Labour Office (1988).[7]

1982, while in Kenya, Tanzania and Zambia they dropped by more than 40 per cent from 1972 to 1983–85. Of the countries shown, only in Zimbabwe have real wages gone up since the start of the 1970s, but even there they fell 20 per cent in just two years, from 1982 to 1984.

The information on woodfuel prices which we turn to shortly must be seen in this context. Although firewood and charcoal prices (in real terms) have remained steady or even fallen in several cities – to the confusion of theories that they should have risen as deforestation worsens – clearly even a falling price can mean greater hardship for consumers whose wages are declining sharply.

The extent of this "fuel hardship" is also extremely hard to measure given the blurring of formal and informal employment in so many African cities. The woodfuel literature is rife with claims that in such and such a city poorer families are spending 20 to 30 per cent – or even more – of their incomes on fuel and are consequently having to cut down on hot meals and other essential expenditures. At a recent UNDP/World Bank seminar on household energy in Eastern and Southern Africa held in 1988, for instance, claims were made that in Zimbabwe the figure could reach as much as 50 per cent and in Addis Ababa over 60 per cent.[8] While some degree of real fuel hardship certainly exists – as it does in richer Northern countries – some of the high claims for African cities invite a good deal of scepticism. Is *food* so cheap in Addis Ababa?

The trouble with nearly all these measures is that they are usually based: (1) on reported incomes which, as we have seen, often have little meaning; and (2) on the assumption that all of a family's measured fuel consumption is purchased. As noted above, the urban poor in particular can often obtain firewood (though not other fuels) without paying for it. Again, this point highlights the need for woodfuel planning to be based on a better understanding of how the poor live and cope with their daily problems.

Urban Dominance: A Cost or Benefit?

As noted earlier, the economic and political power of the larger "primary" cities means that urban populations are able to dominate and outbid rural people in any competition for resources. Since their rapid growth is also a reflection of this power, they will tend increasingly to do so unless macro-economic policies change radically in favour of rural areas.

The enormous distances from which the large cities draw their resources are one sign of this dominance: charcoal, for example, reaches many African cities from as much as 200–400 kilometres away. The widening rings of deforestation in the peri-urban hinterlands of many cities are another. A third is that most cities avoid paying the replacement or environmental costs for their woodfuel where supply involves destructive deforestation.

Since in the larger cities the woodfuel trade amounts to a multimillion-dollar business,[9] all this can be seen as a huge subsidy for urban dwellers which is paid by the environment and by rural people who face increasing competition for their previously "free" wood resources. However, these apparently harmful impacts are not all one-sided.

Consider, for instance, the long supply distances. If charcoal is trucked 400 kilometres to a city, the area from which it can draw its resources is up to 500,000 square kilometres. Except in the most arid regions, such a vast area will provide ample resources and do little harm to them – except possibly in localized patches or strips alongside the main roads. Gerald Foley, for instance, has used this argument to show that Bamako, the capital of Mali – where forests are hardly plentiful – could double its firewood consumption merely by extending the transport distance along the six highways into the city from 175 to 250 kilometres while taking wood from just 7.7 per cent of the land inside this larger area.[10] There would be no need for large afforestation projects, since most of the wood could come from fallow farming areas or natural woodlands which are in any case being cleared for farming.

The "rings of destruction" around cities may indeed be caused to some extent by urban woodfuel cutters but are also due to more basic and inevitable consequences of urban growth. Land within easy reach of urban centres almost invariably provides higher incomes when used to produce food for the city than if left as a forest. Short of deliberate policies to plant or preserve green belts, as the city grows and land values around it increase, woodlands will inevitably give way to more economically attractive land uses.

The impacts of urban woodfuel markets on rural people must also be considered in the round. As we shall see later, wood harvesting and charcoal making are frequently vital sources of additional income and employment for rural populations, while selling to the city trees which they have cut to clear land is simply an added income bonus. More generally, the commodification of rural wood resources is only one aspect of a much more fundamental and, in the long term, inevitable and advantageous transition from subsistence production to a cash economy.

GOALS FOR URBAN ENERGY INITIATIVES

Drawing these points together, one can define many objectives for urban woodfuels and related energy initiatives, each with a range of possible means of approach. We list them briefly here as a convenient reference and as a guide to the places in this part of the book where they are discussed further.

Consumer Welfare

For the mass of poor urban people, energy problems are largely a consequence of poverty and are therefore matters of welfare and distributional justice. The key goals for woodfuel-related initiatives are to:

1. reduce energy costs where they are high or prevent them rising to high levels. This can be achieved by: (a) increasing the efficiency of energy use, e.g. with improved cookstoves. However, interest in improved stoves includes many factors other than energy consumption (Chapter 9); (b) reducing the costs and mark-ups of woodfuel transport and distribution (this chapter); (c) helping consumers to by-pass the woodfuel market, e.g. by forming co-

operatives to buy and truck wood themselves (this chapter); (d) encouraging low-cost woodfuel production (see Supply and Resources, below)
2. improve fuel security, especially the reliable supply and range of accessible fuels for the poor. There are many direct and indirect options for doing this (Chapter 8)
3. improve consumer satisfaction by encouraging the transition away from woodfuels to modern energy sources (Chapter 8).

Supply and Resources: Deforestation

Where woodfuels come from the direct cutting of forests for fuel, or agricultural land clearance, there are many possible initiatives (Chapter 7). Some of them involve greater transport distances or larger payments to resource owners, so that market factors are critical (this chapter). The five main classes of initiative are to:

4. reduce destructive impacts by controlling the scale, methods and location of tree cutting for fuel
5. guard public forests and charge (more) for their use; there are many difficulties in doing this
6. give (partial) control of the forests and revenue benefits from their use to local people
7. increase the price paid for trees felled during land clearance or assist woodfuel traders in reaching felling sites so that more trees are sold rather than burned on site
8. improve the efficiency of charcoal making in order to use less wood per unit of energy produced.

Supply and Resources: Afforestation

There is a similarly wide range of forestry options for increasing woodfuel supplies on a sustainable basis, often with important employment and income benefits (Chapter 7). Again, the price structure of markets and the distances they can reach economically are often crucial. The main classes of option are:

9. natural forest management to enhance production overall or for particular products (e.g. woodfuels), including greater use of forestry wastes

10. "large-scale" afforestation by state or other formal agencies: e.g. peri-urban plantations
11. smallholder tree growing, with woodfuel sales as a principal objective or as a by-product of other management objectives.

Information for Planning

12. improve almost every type of data on energy demand, supply, prices, markets and resources. This point underlies all the others, since the database for understanding and diagnosing woodfuel and related energy problems – especially with regard to the dynamics over time – is with very few exceptions, appallingly weak.

SOARING PRICES?

Pricing is obviously crucial both to urban wood energy consumption and resource use in a number of ways. On the one hand, fuel prices in the city are important to the welfare of energy consumers, how much fuel they use, and their interest in saving energy or switching from woodfuels to modern energy sources. On the supply side, the prices that are paid to firewood and charcoal producers – or the level of taxes they must pay for cutting natural woodlands – has much to do with the adequacy of woodfuel supplies and whether they are "mined" from the forest or grown on a sustainable basis.

This section examines what is known about urban woodfuel prices in Africa and elsewhere, their trends, and the implications for consumers and producers. The next section looks at the market systems which largely determine woodfuel prices.

Received wisdom has it that woodfuel prices in Third World cities have been rising steadily as forests are depleted and transport distances increase. Naturally enough, this has raised many alarms that as fuel prices climb, millions of poorer families will suffer severe hardships. As recently as mid 1987 the FAO, the World Resources Institute, the World Bank and the UN Environment Programme jointly voiced this concern when they warned that: "Shortages (and the distances over which fuelwood must be transported) have caused prices to rise so sharply in recent years that the wood used for cooking often costs more than the food cooked."[11] This exaggerated view has gained such a wide hold and appears so plausible at first sight that it is worth examining briefly.

In a recent review of urban fuel prices in nineteen developing countries, Douglas Barnes of the World Bank[1] made the seemingly reasonable proposal that urban woodfuel prices respond to increasing scarcity in a three-stage process.

In Stage 1, population densities are low and there are plenty of trees near the city. Surplus wood is readily available from nearby land which is being cleared for farming or from regrowth on fallow land. Short distances mean low wood transport costs, there is no sense of anticipation of "scarcity", and so woodfuels are cheap.

In Stage 2, the pressures of population growth on tree resources begin to be felt. There is still plenty of surplus wood from land clearances but it is worth selling this to urban markets only from areas which are close to main roads (or railways). The pressures on these areas increase, new areas must be exploited at greater distances, and transport costs rise. A switch to charcoal may begin at this stage. This reduces unit transport costs and greatly expands the area from which resources can be drawn economically, but further intensifies the pressures on tree resources. For all these reasons, woodfuel prices in the city increase substantially and may rise very quickly as preferred tree species literally "run out".

In Stage 3 there is severe deforestation, woodfuels have to be hauled over long distances, and city prices are high. However, because consumers can switch to alternative fuels, woodfuel prices are "capped" by the prices of these alternatives. The woodfuel trade is forced to trim its operations and margins to this ceiling. Woodfuel prices remain high but can rise, steady or fall because of these adjustments and changes in the prices of competing fuels.

Barnes has tested this model on woodfuel prices and other data over several years for nineteen countries and claims broadly to have confirmed it. The range of prices and rates of price change in his sample were very large. For example, firewood prices varied from US$12 per cubic metre in Sri Lanka to US$248 per cubic metre in the Ivory Coast. Charcoal prices were US$90 per tonne in Madagascar and Zambia but US$670 per tonne in Togo. However, his data were extremely aggregated and, as a result, almost certainly misrepresented the actual conditions pertaining in each of his sample cities. For example, the analysis used all-country averages of national income to estimate urban per capita incomes; and it used the total forest area in each country to reflect the tree stocks within the economic reach of each city.

More importantly, this kind of price-scarcity model neither squares with what appears to have been happening to woodfuel prices in many

Table 6.2: Firewood Price Trends in African Cities

	steady, falling (or erratic)					rising		
year	Botswana Gaborone pula/100kg	Cameroon Yaoundé fr/stere	Ivory Coast Abidjan fr/kg	Malawi Blantyre tamb/kg	Sudan Khartoum sh/kg	Ethiopia Addis Ababa cent/kg	Madagascar Tananarive fr/kg	Zimbabwe Harare cent/kg
1970						4.2		
1971		2,887				4.3		
1972		3,428		2.8		4.9	6.5	
1973		4,099				4.8	6.2	
1974	1.96	4,075				6.3	6.2	
1975		3,937				8.1	7.3	
1976	4.70	2,572	17.9	2.5		7.9	7.0	
1977		3,038	20.1		7.1	10.6	7.0	
1978		4,288	23.1	4.2	5.0	9.9	7.3	1.9
1979		3,988	16.4	4.0		11.1	8.2	
1980	4.30	5,218	17.0	4.0		10.2	8.7	3.6
1981	3.44	6,022	15.0	2.9	4.5	12.4	10.7	
1982	2.55	5,553	14.8		4.3		10.3	
1983	4.83	4,265	14.0	3.0		14.3	9.3	
1984		4,033	13.4	3.5	3.3		9.8	3.7
1985			13.2					3.7
1986			15.7					5.2

Notes on Tables 6.2 and 6.3:
Dir = dirham, fr = CFA franc, kw = kwacha, sh = shilling, tamb = tambala (1/100th kwacha), stere = stacked cubic metre.

Sources: Nominal woodfuel prices: FAO (1983),[13] FAO (1987).[14,15] Consumer price indices: International Monetary Fund (1987).[16] Sudan charcoal: Dewees (1987).[17]

cities nor begins to explain the many social and economic mechanisms which help to account for fuel prices.

Tables 6.2 and 6.3 present all the firewood and charcoal price trends which we have been able to find for African cities and which we believe to be at least reasonably reliable. All the prices are in constant currency or "real terms"; that is, they are corrected for inflation using the local Consumer Price Index normalized to the year 1980. In other words, the prices are in local currency of 1980 purchasing power. This fact explains apparent inconsistencies between the prices presented here (in 1980 currencies) and elsewhere in the chapter where prices are for different years. Prices have not been converted to a convenient, common unit such as the US dollar because this would introduce gross distortions across the countries arising from artificially high or low foreign-exchange rates. However, because the price trends are not easy to read from the tables, Figures 6.1 to 6.4 present the trends for the twelve cities for which there are long-term records, with prices in the first three tables converted into US$ (1980) per kilogram.

Before we discuss the trends, a word of caution is necessary. Obviously, the data are only as good as the statistics on which they are

Table 6.3: Charcoal Price Trends in African Cities

year	\	steady, falling (or erratic)							steep rises			
	Burundi Bujumbura fr/kg	Cameroon Yaoundé fr/kg	Ghana (urban) cedi/kg	Madagascar Tananarive fr/kg	Morocco Casablanca dir/kg	Senegal Dakar fr/kg	Sudan Khartoum S£/100kg	Zambia Lusaka kw/bag	Ethiopia Addis Ababa cent/kg	Ivory Coast Abidjan fr/kg	Kenya Nairobi sh/kg	Tanzania Dar es Salaam sh/kg
1970						31.4			40.4			
1971		92.1		19.9		30.2			41.5			
1972		97.3		19.3		30.8			47.1			
1973		81.5		18.3		27.7			45.9			
1974		79.0		17.7		32.9			60.2			0.72
1975	17.1	71.2		15.2	0.95	27.7	7.2		89.6			0.95
1976	17.0	79.8		15.6	1.02	27.5	8.2		74.0		0.63	0.95
1977		77.4		16.3	1.04	24.7	8.7		76.6	39.3	0.64	1.50
1978	20.3	93.5		19.5	1.18	23.9	8.5		93.1	52.1	0.59	2.02
1979	19.8	90.9	1.25	18.9	1.31	21.7	9.0	3.4	95.5	56.2	0.54	1.43
1980	18.5	113.0	1.25	16.6	1.2	20.0	9.2		103.6	71.0	0.70	1.25
1981	21.6	87.6	1.14	21.0	1.16	27.4	9.1		95.2	68.9	0.71	1.47
1982		93.3	1.05	19.5	1.05	23.3	8.9		101.8	65.9	0.85	2.72
1983		91.6	1.01	22.9	1.14	20.9	9.3	2.7		60.6	1.02	3.34
1984		79.8	1.21	21.9	1.01	18.7	8.5	4.3		58.1	0.98	2.47
1985			1.18	24.4	1.13	16.6		3.5		67.0	0.90	1.87
1986			1.03			21.5		3.3		71.4	0.81	1.74

Source: As Table 6.2.

Figure 6.1: Firewood Price Trends in Four African Cities

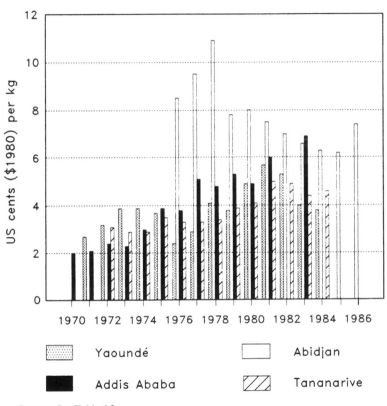

Sources: See Table 6.2.

Figure 6.2: Charcoal Price Trends in Four African Cities: Steady Prices

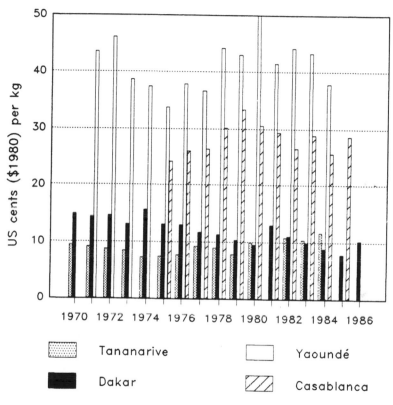

Sources: See Table 6.2.

Figure 6.3: Charcoal Price Trends in Four African Cities: Rising Prices

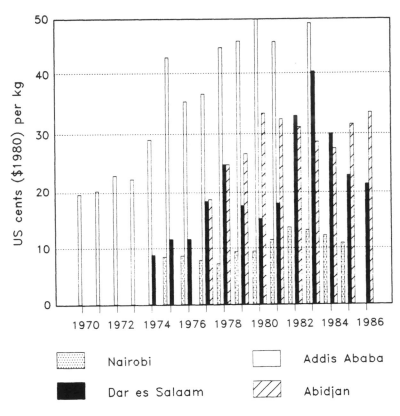

Sources: See Table 6.2.

Figure 6.4: Charcoal Price Trends in Khartoum (Sudan)

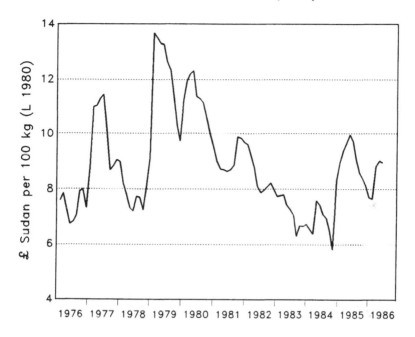

Source: Dewees (1987).[17]

based. Urban commodity price statistics are not always reliable, especially when there are large seasonal price variations, or large price variations across city markets, or differences between official and actual market prices – as is often the case with woodfuels. For similar reasons, consumer price indices which record the combined price of a defined basket of goods and services, may be unreliable – especially for particular social groups.

Nevertheless, we believe that these price data are the most comprehensive and reliable that are currently available and that any errors in them are not so serious as to invalidate the broad conclusions that can be drawn from them. Generally speaking, the data are probably most reliable for the cities with continuous records because these have well-established methods of monitoring retail prices.

If we consider only these cities and use regression analysis to smooth out the price fluctuations, three main types of price movement are evident:

1. *Long-term reductions.* Firewood prices fell by about 3.5 per cent a year in Abidjan, Ivory Coast, a country which has experienced some of the highest rates of deforestation in Africa.[1] In the seven years prior to 1984, they more than halved in Khartoum (Sudan) where deforestation has also been severe. As for charcoal, prices fell in Dakar (Senegal) by 3 per cent yearly over the long period from 1970 to 1986, while there was a small but fairly steady fall in "typical" urban prices in Ghana.
2. *Long-term stability,* notably for charcoal in Yaoundé (Cameroon), Casablanca (Morocco), Khartoum (Sudan), and Ghana.
3. *Sharp and short-term rises, followed by steady or falling prices.* For firewood, the most striking examples are Yaoundé (Cameroon), where prices roughly doubled in real terms between 1976–7 and 1981 but then fell by a third from 1981 to 1984, and Addis Ababa (Ethiopia), where they more than doubled from 1973 to 1977 and then rose sharply. For charcoal, all four cities shown in the right half of Table 6.3 have experienced extremly sharp price rises for a short time. In Addis Ababa, Abidjan, Nairobi and Dar es Salaam prices roughly doubled in the space of only three to four years, but have since tended to steady or fall.

These latter cities do seem to have followed the classic three-stage price-scarcity model outlined on p.197. However, if we look more closely into these trends some curious points emerge.

One obvious feature of the data is that in most cities there are quite large but short-run changes from year to year. At least some of these are known to be real and not due to statistical errors. For example, in Nairobi charcoal prices shot up by 20 per cent between 1981 and 1982, causing much public protest, due to temporary shortages of kerosene and bottled gas.

Even more striking are the cases where long-term quarterly or monthly price data are available. The most interesting example is Khartoum, where deforestation is extreme and charcoal is now hauled from as much as 400 kilometres away. Although one might expect these long haulage distances to dominate prices and produce a generally rising trend, in fact prices have fluctuated rapidly, with inter-seasonal differences reaching nearly 100 per cent during the dozen years for which records are available, but they have not risen overall. Some of this variation could be due to errors in the inflation index which is used to convert current prices to prices in real terms, but this is hardly likely to explain all of the remarkable variations. The charcoal price trends for Khartoum are presented in Figure 6.4 (p.203) and discussed in Urban Case 3.

Equally telling evidence that woodfuel prices need not follow the simple scarcity model and are only loosely related to levels of deforestation comes from studies for India. Urban firewood price trends for some forty Indian cities (corrected for inflation) vary widely. For instance, from 1970 to 1982 they rose by 482 per cent in the state of Uttar Pradesh and 121 per cent in Orissa, but altered by only 2 per cent in Haryana, and fell by 35 per cent in Maharashtra.[18] These changes bear little relation to the areas of closed forests plus open woodlands which have been established recently by the use of space satellites for vegetation mapping – which, incidentally, showed that in India as a whole deforestation rates from 1972 to 1982 were seven times faster than previously supposed. For example, the largest price falls were in Rajasthan and Maharashtra, which ranked third and seventh in terms of forest loss out of the twenty-one major states, with area reductions of 47 and 25 per cent respectively. In Haryana, which ranked second with a 50 per cent forest loss, urban prices hardly altered.

A possibly unique study tried to resolve these discrepancies by comparing firewood prices with satellite data on the forest area in a circle of 100 kilometres radius around each city.[18,19] The cities ranged from Bombay, Calcutta and Delhi with populations of over 5 million to several with fewer than 300,000 inhabitants, while the forest resources varied from 9 hectares to one-thousandth of a hectare per city dweller.

The study did find a clear if rather weak relation between firewood prices and the forest area per person: as the latter decreased by a factor of nearly 10,000 firewood prices rose by a factor of four to five. However, against all expectations, the largest price increases occurred in the cities with the *greatest* forest resource per person, while there were almost threefold price differences in cities with much the same per capita forest resources around them. Also, the main influence on prices was not so much the size of the forest resource but the size of the city. Clearly some other important factors were at work. Those factors are the woodfuel traders.

WOODFUEL MARKETS

Woodfuel markets can be very complex and dynamic, with many stages and actors along the chain from wood harvesting or charcoal making at one end, through transport to and in the city, to the great variety of wholesalers and retailers who sell to consumers. Even though these trading networks often amount to multimillion-dollar businesses, remarkably little is known about them. They have been studied carefully for only a few cities, and even then rarely for more than one period of time.

Furthermore, studies which appear thorough at first sight often turn out to contain rough estimates or leave out costs which can make all the difference to one's understanding of how efficiently the market is working, how well people are doing in the business, and how the market might adjust to changes in supply and demand conditions. One of the commonest faults is in the treatment of gross margins or "mark-ups": for example, the difference between a retailer's buying and selling price for firewood. Major cost items such as labour for splitting wood may be ignored, thus falsely inflating the portion of the mark-up which constitutes the retailer's income (itself often referred to as "profit").

Similarly, large mark-ups of several hundred per cent, even after all costs have been deducted, are frequently said to show that traders are making huge incomes when in fact, because they sell so little each day, their incomes are quite modest. This is particularly true of the scores of "petty traders" operating from home and providing a vital service to local families who cannot easily reach the larger central markets. For instance, a small and informal survey by one of the authors in Blantyre in 1988 found that traders who were selling firewood at around three times their total costs were selling so little each day that their net incomes were only just above the minimum urban wage and never exceeded that of a truck driver.

Nevertheless, enough is known about these distribution and market systems to establish two crucial points: (1) that costs and margins can be extremely variable; and (2) that there is therefore usually large scope for reducing costs or otherwise increasing the economic efficiency of the market – whether as a result of natural forces or designed interventions.

Such changes can make all the difference to urban woodfuel supply, its impact on tree resources and the prospects for sustainable wood production for urban markets.

Quite simply, if unit costs and mark-ups can be reduced then raising retail prices – an important consideration for poor consumers – the woodfuel market can either pay more for tree resources or reach further for them, or both. Paying more improves the economics of woodfuel production, whether in formal plantations or smallholder woodlots, and also the likelihood that salvage wood from land clearance operations will be sold and not burned. A greater market reach allows the city to tap more distant and possibly surplus natural forest resources and, of course, enlarges the range over which wood production for the market is profitable.

Market Structures

Figure 6.5 presents a simplified diagram of the possible variations of woodfuel transport and distribution networks and shows that there may be anything from one to five (or more) stages between the primary producers and the final consumer. This section presents a brief overview of the main stages in the network and the key points of note about them. These are discussed further in later sections using particular examples.

Resource prices

With natural forests, most countries attempt to protect state property from illegal woodfuel cutters by licensing some cutters for a fee, or charging woodfuel transporters on their way to town. Evasion of such "stumpage royalties" is normally easy, while the incentive to avoid them is stronger the higher the taxes. If the forest is under the legal or customary control of local villagers there is probably a better chance of preventing illegal cutting and therefore of raising some revenue from use of the resource. At present, such revenues – if paid – are rarely more

Figure 6.5: Possible Stages and Supply Chains for Woodfuel Markets

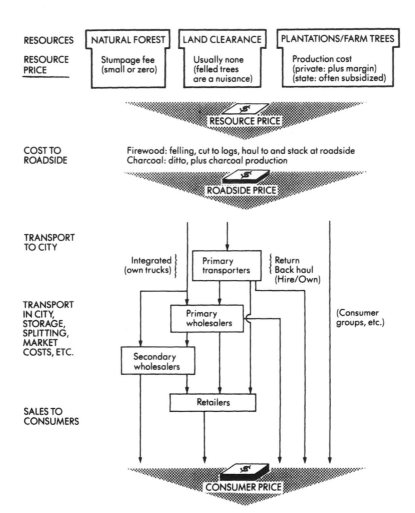

than 5 per cent of the final consumer price. Similarly, with land clearance and slash and burn farming systems it is wholly impracticable to raise resource taxes unless one is dealing with large, commercial operations which can easily be identified in advance. We return to the difficulties and advantages of guarding and taxing forest resources in Chapter 7.

By and large, only with government or private plantations (or small-holder tree growing) can a price be charged which covers the costs of production or, in other words, the replacement cost. However, several governments subsidize sales of wood for use as fuel on the grounds that it is a basic need and prices must be held down. For example, around Blantyre, Malawi, the Forest Department sells stacked wood in their plantations for 2.9 kwacha per (solid) cubic metre if the logs are less than 2.5 metres long. Above this length the logs are presumed to be for timber or building and telephone poles rather than firewood, and prices range from 14 to 34 kwacha per cubic metre depending on their diameter and length.

Roadside prices
Like the rest of the woodfuel distribution system, felling trees and cutting them into logs of manageable size, hauling them to the roadside and stacking them is labour intensive. It may also involve substantial capital investments for axes, saws, barrows, carts, etc. For example, a World Bank survey near Dakar (Senegal) found that it can take as much as sixty man-days to cut, carry, stack and load a medium-sized truck carrying 30 (stacked) cubic metres of large logs and branches weighing 12 tonnes (see Table 6.6). Even with a daily wage as low as 350 CFA francs (US$1.3) this works out at US$77 to load the truck. In Blantyre, on the other hand, the Forest Department expects three times the work rate – that is, for one man to cut, haul and stack 1.5 cubic metres in a day. One man-day per stacked cubic metre is a common norm for commercial plantations.

Rural wage rates and labour efficiency and supervision (or motivation) are thus key influences on prices. This is particularly the case for charcoal making, which is even more labour intensive than harvesting firewood and is usually carried out by migrant workers or small, part-time operators who are seeking to supplement meagre incomes from farming. The small charcoal makers who dominate production in most African countries therefore tend to show little interest in attempts to improve their use of tree resources – notably, by adopting more efficient charcoal kilns – unless these produce more quickly. Squeezed between his high labour costs and the price which the truckers from the city will pay for his charcoal, his primary interests are in saving time, not saving forests.

Transport costs

These are normally a major part of final woodfuel prices but depend on many factors apart from simply haulage distance. These factors include truck size and payload, road conditions as they affect vehicle speeds, and fuel and repair costs. In many African countries, operational trucks are so scarce that owners can raise their charges to those of the highest bidder, which is usually someone who carries high-value or perishable goods: that is, many things other than woodfuels. This can lead to large seasonal and annual cost variations for hauling woodfuels. The time needed to pick up a load of wood can also be a critical factor in labour costs and can give a significant economic advantage to supplies from natural forests and plantations compared with those from dispersed smallholder producers.

Modes of transport operation are also critical to costs and thus to overall price mark-ups. The main difference is between *return transporters* who deal only with woodfuels and make an empty journey from the city to the rural pick-up points, and *back haulers* who carry loads on the outward journey and woodfuel on the return trip to the city, or simply pick up some woodfuel on the last leg of a general haulage trip to various urban centres. With back hauling the costs of transporting woodfuels are considerably smaller than for return transporting.

The switch from return transporting (typical of present-day Kenya, Malawi, Tanzania and Zambia, for example) to back hauling (typical of the highly organized charcoal trade in Sudan) is one of the largest cost-reducing measures that can be made as pressures on the market system tighten. A third stage is *integrated transporting* in which trucks are owned and operated by high-volume wholesalers who thereby gain better control over supplies and price operating margins.

Vertical integration

Integration of this kind between the various woodfuel production and distribution stages is another major route to lower overall costs and mark-ups. For example, wholesalers might operate their own transport and hire labour to harvest woodfuels or make charcoal, thus reducing total overheads or allowing losses in one operation to be covered by profits in another. This has occurred in some countries, such as the Sudan (see Urban Case 3). The ultimate in vertical integration is shown on the right-hand side of Figure 6.5 (p.208), where a consumer co-operative, for example, cuts out the middle men by buying directly from the producers, hiring and driving its own vehicle which delivers "to the door". With firewood, member households would do their own chopping to smaller-size pieces, avoiding the large "break of bulk" costs normally borne by the retailers.

Wholesalers and retailers

The final stages in the market chain provide numerous services, each with various direct and overhead costs. These services include transport within the city, "break of bulk" to small volumes for sale, and bagging and bundling. Storage and security costs and stall licence fees may be considerable. Charcoal retailers who sell by the tin or in other small volumes must also bear the cost of losses of saleable charcoal – often of around 15 per cent – due to the breakdown of lump charcoal in the sack to powder or "fines". Fuelwood market structures are usually more complex than for charcoal, since there is typically a greater mix of transport modes and operators (foot, donkey-cart, small to large truck, etc.) and also of firewood qualities and unit volumes for sale.

Break of bulk

This is one of the major services provided by the market and – like any other distribution system – is mainly to benefit the mass of consumers who can neither afford to buy nor are able to carry fuels in anything but small amounts at a time. Wood logs off the truck must be chopped into small pieces while charcoal must be distributed from the sack into small tins or piles on the floor of the market stall. Chopping logs by hand – often with poorly made and blunt axes – is hard, time-consuming work, but again the labour costs involved can vary enormously from one market or wood seller to another. As a result, small volumes of firewood and charcoal typically cost roughly twice as much per unit weight as large logs or charcoal sacks weighing 35–45 kilograms. The same twofold price difference is also found for small-volume kerosene sales in local markets compared to the pump price in, for example, garage forecourts.[20]

Table 6.4: Price Structures for Urban Firewood Markets

	Burkina Faso Ouagadougou 1983	Chad Ndjamena 1976	Mali Bamako 1983	Niger Niamey 1984	Senegal Dakar 1981	Kenya Nairobi 1986
Retail price	100	100	100	100	100	100
Distribution	22.7	64	17.2	20.0	23.2	21.7–33.7
costs	—	—	2.0	6.7	14.1	16.9–13.5
income	—	—	15.2	13.3	9.1	4.8–20.2
Ex-truck	77.3	36	82.8	80.0	76.8	78.3–66.3
Transport	31.8	29	31.6	33.3	31.7	31.4–33.7
costs	27.3	20	20.8	26.7	—	19.3–20.2
income	4.5	9	10.8	6.6	—	12.1–13.5
Price to producer	45.5	7	51.2	46.7	45.1	46.9–32.6
costs	—	—	6.4	35.3	8.8	8.2–5.7
income	—	—	42.8	11.4	34.7	29.5–20.5
taxes	0	0	2.0	0	1.6	9.2–6.4

Note: Income may be inflated because of unidentified costs.

Sources: Burkina, Chad, Mali, Niger: Bertrand (1986).[21] Senegal: UNDP/World Bank (1983).[22] Kenya: Dewees (1985).[23]

Examples of Production and Market Systems

Tables 6.4 (p.210) and 6.5 present the price structures of firewood and charcoal markets in some African cities. Perhaps the most striking points about them are, first, the large variation from one city to the next and, second, the paucity of information provided by the original sources for these data. The most serious lack of information is on how the mark-ups at each stage break down into costs for materials, into hired labour, and into income for the main operators. There is thus no way of telling whether traders are making rich or meagre livings out of their work.

With the firewood markets, the prices and margins at each major stage are fairly similar. With the exception of Chad and the right-hand column for Kenya, the roadside price is 45–50 per cent of the final retail price, transport accounts for 30–35 per cent, and wholesaling and retailing add the last 20 per cent or so. The charcoal markets, on the other hand, are much more variable. Most noticeably, in Mozambique and Senegal final distribution accounts for about half of the retail price, while in Madagascar and Tanzania it makes up only 10–20 per cent.

Two other points are worth noting. First, the distribution mark-ups

Table 6.5: Price Structures for Urban Charcoal Markets

	Madagascar Moramanga 1987	Mozambique Maputo 1985	Senegal Dakar 1987	Tanzania Dar es Salaam 1986
Retail price (bag)	100	100	100	100
Distribution	20	47.6	50.0	11.1
costs	7.5	—	—	—
income	12.5	—	—	—
Ex-truck	80	52.4	50.0	88.9
Transport	30	7.7	25.3	51.9
costs	12.5	—	—	—
income	17.5	—	—	30.2
Price to charcoaler	50	44.7	22.2	36.9
costs	13.5	—	—	—
income	25.5	—	—	—
taxes	1.0	0	2.5	3.9

Note: Income may be inflated because of unidentified costs.

Sources: Madagascar, Senegal: Matly (1987).[24] Mozambique: UNDP/World Bank (1987).[25] Tanzania: UNDP/World Bank (1987).[26]

are very much higher than for modern fuels, reflecting the amount of labour needed to "bulk break" and prepare woodfuels for sale and the small-scale, dispersed nature of woodfuel selling. Whereas the final distribution mark-ups for woodfuels are mostly in the 20–50 per cent range, World Bank surveys of kerosene market structures in seven African cities found distribution mark-ups of only 3 to 7 per cent.

Second, in no case would the roadside prices begin to cover production costs from plantations. Instead, they mostly reflect labour costs for tree felling and stacking. In only three cases are any stumpage royalties ("taxes") paid, amounting to only 2 per cent of the final price in Mali and 6 to 9 per cent in Kenya.

Table 6.6 gives a breakdown of firewood production and marketing in Dakar (Senegal), and amplifies several key points about the general nature of woodfuel markets. First is their high labour intensity. The labour items which the survey was able to identify (and as the question marks indicate, these were by no means all) amount to just over six man-days for a solid cubic metre or 400 kilograms of firewood, including three for harvesting and two for splitting. That quantity of wood, air dried, has almost exactly the same energy content as a barrel of oil. If we apply the present average UK industrial wage of around

Table 6.6: Price Structure and Labour Costs for Fuelwood Distribution to Dakar (Senegal) 1981

| | per solid cubic metre | |
	man-days	francs
Retail price (fr. 20 per kg)		12,800
Chopping into saleable pieces (urban wage fr. 800 (US$3.3) a day)	2	1,600
Unloading from truck (urban wage)	0.13	210
Retailer margin (*assumed*) (actual, including other costs)	??	1,164 ??
Transport, Koumpentoum-Dakar fr. 70,000 (US$258) for 30-stere load (labour, other costs)	??	3,733 ??
Loading truck (rural wage fr. 350 (US$1.3) a day)	0.9	320
Cutting, stacking (rural wage)	3.2	1,120
Permit fee (fr 250 per stere)		200
Residual to include costs marked ?? above and, according to the World Bank, a "rent" to reflect scarcity value of wood		4,453

Notes: 1 stere = 0.625 solid cubic metres, or 400 kg. CFA francs 272 = US$.
Source: UNDP/World Bank (1983).[22]

US$70 a day just to the cutting and stacking stage, the labour costs for these operations alone would be about $210 per barrel of oil equivalent. This compares with the actual total oil production costs in the Middle East of less than US$2 a barrel.

Woodfuel prices are therefore extremely sensitive to the costs of labour and are likely to come under severe pressures to reform and streamline themselves as – or if – wages rise.

A second and related point is that urban prices are often held down by low returns and labour wages in rural areas. In the case of Dakar the rural wage for tree felling and stacking was well under half the urban wage for splitting wood and other marketing tasks. If all the rural operations had been performed at the urban wage rate, the Dakar price would have been at least 15 per cent higher.

Similarly, studies by the World Bank on woodfuel supply options for Nairobi found that in 1986 labour rates near the city were 30 Kenyan shillings per day. But they were only 18 shillings for rural labour and 11 shillings for migrant rural workers who are frequently involved in casual wood harvesting and charcoaling.[20,23] If new production projects such as peri-urban woodfuel plantations or organized charcoal making sites are placed close to the city rather than in more remote rural areas, lower transport costs can be more than offset by higher labour costs. These relativities will also change over time as the city expands and wage rates and land prices increase in widening belts around the city.

These distance/cost trade-offs are not well understood, but will obviously be specific to each city and can also change rapidly both seasonally and year to year. This underlines the point made earlier that the diversity and dynamism of woodfuel markets make generalizations about them difficult; but these factors also make them far more interesting and malleable than simple price-scarcity models would have us suppose.

Thirdly, Table 6.6 shows that woodfuel trading on this scale is not a poor man's game. Large outlays and capital costs, as well as financial risks, are involved: at 70,000 CFA francs, the transport costs for one truckload amount to nearly a year's wages for a rural worker. For a small commercial trader such costs and attendant risks can be ruinous and help to account for high woodfuel prices in some places. For example, one of the authors heard repeated complaints from small traders in Blantyre, Malawi, that they would often hire a truck and loading crew to fetch wood from a sawmill or Forest Department site only to find when they arrived that there was no wood for sale.

To conclude this section, Urban Case 1 looks at a woodfuel production and transport system at the opposite pole to the one discussed on p.213 for Dakar. Based on a survey of most of the firewood cutters and transporters operating around Gaborone, the capital of Botswana, it is a story of donkey-carts rather than 30-ton trucks and of people who engage in the woodfuel trade not to get rich quickly but as a way of surviving the poverty and hardships of rural life. One of the great but critical unknowns about the woodfuel markets of Africa is how much of their activities are conducted by people such as these, how much by traders with greater assets and commercial resilience, or at the other extreme how much by the "big boys" who monopolize the trade in some towns and cities. What can be done to streamline the economics of woodfuel markets so that they provide a better match between sustainable resource use and the needs of poor consumers depends crucially on this question.

TOWARDS MORE EFFICIENT MARKETS

In this section we present three case histories of actual woodfuel markets to highlight the key factors other than "wood scarcity" which determine final prices and to show how markets can change to become more economically efficient. Such changes can give pointers to the kinds of interventions that could help to accelerate this process.

Tanzania: Towards Integration

Urban Case 2 presents a recent picture of charcoal production and trading in three cities in Tanzania during the wet and dry seasons. One major point revealed by this case is the large inter-city and seasonal variations in costs and mark-ups. In particular, in the wet season production costs are higher because wood felling, drying and charcoal making all take longer. Transport costs may be double those of the dry season because of shortages of trucks, spare parts and fuel. Easing these shortages or controlling "excessive" truck hire costs could do much to reduce transport costs.

A second key point is that much of the trade is beginning to become well organized and integrated; urban prices are lower where this is the case. However, whereas prices paid to producers are relatively high and final distribution mark-ups are low (see Table 6.5, p.213), there is a good deal of return transporting in which trucks are empty for the trip from the city to the charcoal pick-up point. This suggests that there is still considerable scope for reducing the relatively high transport costs which make up about half the final price.

Competition in markets with many participants should lead to the efficiency improvements noted above, but perhaps only when retail prices are rising fast so that the market is squeezed by consumer resistance to high prices.

Ouagadougou: A Market Reforms

One remarkable example of a market which appears to have made such changes is Ouagadougou (Burkina Faso), where there is information on market price structures for a number of years.

Table 6.7 shows that as firewood and charcoal prices in Ouagadougou rose rapidly in real terms during the last half of the 1970s there was a major shake-out in the market structure, followed by a steep fall in prices. The price mark-ups by transporters were sharply reduced despite rising costs, while the average price paid to fuelwood producers more than doubled – and trebled in terms of the percentage share of the retail price. Large changes also occurred in the final stages of the distribution chain, most notably through a sharp fall in distribution

Table 6.7: Changes in Woodfuel Prices and Cost Components: Ouagadougou (Burkina Faso), 1975–84

	1975	1980	1984
Retail prices			
(1980 CFA francs per kg)			
firewood	11.6	15.0	10.4
charcoal – pile	73.2	58.0	43.0
charcoal – sack	40.8	50.0*	34.0
Firewood price structure			
(1980 CFA francs per kg)			
retail price	11.6	15.0	10.4
distribution	3.3	5.0	0.9
transport	7.1	6.3	5.7
producer price	1.2	3.7	3.2
Firewood price structure			
(per cent of retail price)			
retail price	100	100	100
distribution	28.6	33.3	8.5
transport	61.4	42.0	54.9
(costs)	(21.4)	(25.3)	(36.6)
(margin)	(40.0)	(16.7)	(18.3)
tax	—	—	5.6
producer price	10.0	24.7	31.0

Note:
* Price range was CFA francs 40–60 per kilogram. Sales by sack tend to be wholesale, by small pile retail.

Sources: UNDP/World Bank (1986),[27] Bertrand (1986).[21]

margins and in the price of small volumes of charcoal (which are bought by the poor) compared with charcoal by the sack. These changes were so great that the producer tax of 6 per cent of the final retail price which had been introduced by 1984 did not prevent final prices from falling sharply.

In short, by reducing the costs and mark-ups in the middle levels of the distribution system, these extraordinary changes benefited poor consumers by bringing down woodfuel prices, and helped the producers by paying them more. It will be interesting to see if such beneficial changes can be sustained and, more importantly, if regular market surveys elsewhere can show that similar changes are occurring in other cities.

Sudan: the Professionals

Finally, Urban Case 3 present some information on perhaps the most efficient and highly organized charcoal market in Africa, one which offers many lessons on how markets in other places might evolve – or be helped to change – in order to hold down distribution costs and allow greater revenues to be made from producing or harvesting wood while keeping down prices for the urban poor.

One key feature is that much of the distribution system is highly organized, integrated and professional. Without using high-cost hardware, many of the charcoal makers are highly experienced and often achieve exceptionally high weight-for-weight conversion efficiencies of 25–30 per cent, or around three times better than in Tanzania. Production is estimated to require 15-20 man-days per ton of charcoal compared with 25–30 man-days per ton in other parts of Africa, thus helping to increase charcoaling incomes. A large part of the trade is run by a few entrepreneurs who use sophisticated credit financing schemes and agents in the field who each recruit, supervise, finance and provision some 50 to 100 charcoal makers. Charcoal storage is a common feature which helps to smooth out the higher costs and falls in production during the rainy season.

Charcoal transport is particularly well developed and cost efficient, since most of it is by trucks which pick up a charcoal load on their way back to the city after delivering other goods. Indeed, charcoal carriage is seen as a way of *reducing* costs for general haulage. As a result, and despite enormous haulage distances of up to 400–500 kilometres, transport to the city accounts for less than 20 per cent or so of the final retail price – half as much as in many African cities such as Nairobi (Kenya).

In short, Sudan provides a pointer to the way in which woodfuel markets in Africa may have to develop by modernizing their methods of financing, management and supervision and, through concentration of ownership, by achieving greater economies of scale in their operations. Many small traders will lose out in such a development but many others – not least the mass of consumers and people who would like to produce and sell woodfuels for the market – could gain.

The main lesson of this chapter, however, is that no developments of this kind can be engineered by governments or donor agencies unless they first gain a good understanding of how the woodfuel market in each major urban centre actually operates and where they key constraints and opportunities for improvement lie. Given the complexity of woodfuel markets and our poor understanding of them, that will be a major task.

Chapter 7

Trees for the Cities

The cities of Africa will need more woodfuels, and will need them in enormous volumes. Setting aside the question of whether woodfuel consumption can be reduced, which we examine in Chapter 8, how can urban demands for wood energy, as they are growing at present, be met in adequate and sustainable amounts at affordable prices? That is the question for this chapter.

Schemes which are aimed either at protecting natural forests to ensure woodfuel supplies, or at growing trees for urban woodfuels, have been major features of the forestry and energy strategies of both donor agencies and African governments for the past couple of decades. There is a wide range of possible approaches, some of which have considerable potential for taking pressures off the natural forest, or using them more effectively, or increasing tree resources outside the forests.

However, as with rural tree growing and management, disappointing experiences have led to quite profound changes in approach in the past few years. Classical "industrial" forestry projects – typified by commercial, peri-urban "energy" plantations – have usually failed to meet expectations and have often run into numerous and complex constraints which were never foreseen.[1] Consequently, there has recently been a surge of experimentation on new ideas, essentially of two kinds:

- for natural forest management, a switch from restrictive approaches by state agencies towards promoting forest management by local people who also share some of the cash benefits
- for tree growing, a switch from formal plantation forestry towards "softer", participatory methods of encouraging farmers to grow trees for the market; that is, taking aboard the lessons spelled out in Part I of this book.

This chapter does not attempt a comprehensive review of the great variety of forestry practices and experiences in Africa. Instead, it looks briefly at the major issues and options to do with safeguarding or increasing urban woodfuel supplies and the arguments for adopting these new and more flexible approaches rather than the classic forestry models.

TAXING AND GUARDING THE FORESTS

Deforestation to provide agricultural land and construction materials appears to be an inevitable feature of development. It occurred on a massive scale in Europe and North America and helped to underpin the industrial revolution in these regions by providing a secure agricultural base. Much the same is now happening in Africa, and for similar reasons.

Against this trend, numerous schemes have been tried or proposed for guarding state forests against commercial woodfuel collectors and charcoal makers and/or making them pay for the use of these resources. Indeed, in many African countries the idea of taxing the use of state forests has become a major arm both of forestry and woodfuel strategies and could even become a condition for financial support from some international donor agencies. It has also become a highly controversial idea. While in theory its economic logic appears impeccable, in practice it raises a host of serious difficulties.

The general argument for guarding and taxing forest resources is easily made. Forests are assets with many economic values, whether for commercial or subsistence products, environmental service functions (for example, to protect watersheds and soils, maintain genetic diversity, provide habitats for forest dwellers and wildlife, or serve as carbon sinks which help to reduce global warming from the release of "greenhouse gases"), or aesthetic value. Owners of such resources, whether state or private, have a right to charge for their use. Furthermore, if the state could charge for these resources substantial advantages could be gained. At least in theory:

- state revenues would be increased; these could be directed to improving forest management or the support of tree growing, whether by the Forest Department or smallholders
- forests would be used less wastefully if users had to pay; for example, more trees would be sold rather than burned on site during agricultural land clearing operations, woodfuel cutters

would be more careful about how much they cut, and charcoal making would become less wasteful of wood

- incentives for tree production outside the forests would be increased, since producer prices would rise and there would be less competition with "free" wood taken from the forests.

Moreover, the revenues involved could be very large for poor countries. Just as an illustration, a city of 250,000 people who depend on charcoal may consume something like a million bags of it a year. If the forest wood for making the charcoal was taxed at the same royalty rate as, say, crude oil – or roughly at the equivalent of US$12 a barrel (US$2.1 per GJ of energy) – the tax on a charcoal bag would be US$2.5, or about US$4 to US$6.5 for a cubic metre of tree wood, depending on conversion efficiencies.[2] The revenue would therefore be around US$2.5 million a year. It is worth noting that these seemingly high tax rates are about the same as the costs of government-run plantations in Kenya and Tanzania (see Tables 7.2 and 7.3, pp.233, 234).

However, against this positive view one has to consider the many practical difficulties and social problems that face attempts to collect revenues from forest users.

The first concerns forest clearances for making new agricultural land, and "slash and burn" cultivation systems: the main causes of deforestation in Africa and a major source of rural and urban woodfuel supplies. Where large, commercial farm operations are invovled it should be relatively easy to impose and collect substantial royalties on any trees which are felled. In the Sudan, for example, where some 4.5 million hectares have been cleared for mechanized agriculture and clearance of another 1.7 million hectares is expected by the end of the century,[3] most of the cleared trees are burned in the field because licences to use land are frequently given late and the trees must be disposed of quickly before the planting season begins. If tree cutting were taxed, greater efforts would be made to clear earlier, store the trees, and find markets for them. Given the low rents on the controlled land in these schemes – currently about S£ 2.3 (or just under US$1) per hectare annually[3] – there is little question that substantial resource rents could be applied to the cleared trees without harm to the economics of agricultural production.

But such large and visible operations are the exception. It is inconceivable that similar taxes could be imposed on the myriads of smallholders, most of them subsistence farmers, who clear trees to farm for survival. They could not spare the cash to pay any but the most paltry tax. Nor, if they did pay, would it save many trees. Felled trees in such systems may be important to farming itself, or may be just a nuisance. In the former case, trees are left to rot on the ground to maintain soil fertility, or are burned to clear space for planting and produce ashy nutrients. In favourable places close to roads or near an urban centre, "surplus" trees may well be sold into the market. However, to cut them up and haul them to the roadside for sale requires a lot of labour at just the time when planting is starting and farm labour scarcities are often most acute. Charging a tax on this type of tree clearance would make no difference to these labour constraints, nor to the ability of urban traders to pay more or reach further afield for tree resources.

The second difficulty is widely recognized. How does one catch the woodfuel cutters in order to tax them? Primary resources are easy to tax under two conditions: if there are relatively few visible resource users (as with oil wells or metal mines, for instance) or if there is a well-developed and expensive bureaucracy of tax collectors to reach scattered resource users such as farmers. Neither condition applies to the forests of poor countries, where myriads of resource takers work in scattered and often remote locations while handfuls of forest guards, usually with no more than bicycles (if they are lucky), try to police them. Even in relatively prosperous Brazil, there are only 300 guards in the Forest Defence Service attempting to police the 5.5 million square kilometres of the Amazon basin.[4]

Not surprisingly, attempts to police the primary producers – the woodfuel cutters and charcoal makers – have hardly proved feasible anywhere. Even if the occasional illegal operator is caught it is unlikely to deter the majority as long as their activities are providing essential incomes and livelihood. Furthermore, if taxes are raised substantially, as most governments would like, the incentives not to get caught are simply increased.

But perhaps most importantly, the role of forester as (armed) guard and tax collector does not fit easily with the increasing recognition that forest services must be seen to be the ally of rural people if they are to work with them successfully in tree growing and management. One might also ask whether the strict policing of forests which once belonged to local communities is an acceptable route to sustainable development.

To ease the problems of tax evasion, many governments focus their controls and taxation of woodfuels on the transporters, since it is relatively easy to stop trucks at check-points on a few main roads entering the cities. This has generally been more successful, but perhaps only because the taxes are almost universally small. In the examples of woodfuel price structures presented in Tables 6.4 to 6.7 (pp.210, 217) the highest resource tax as a percentage of the woodfuel's retail price was 9 per cent for charcoal in Kenya; in most of the other countries the equivalent figure was from 1 to 5 per cent. Willem Floor of the World Bank[5] has reported that in West Africa forest taxes are never higher than 5 per cent of the final retail price. Furthermore, such taxes are also easily evaded. Floor estimates that in Niger and Mali only 5 to 10 per cent of woodfuel producers and dealers who are liable for taxes actually pay them, while in Tanzania a large UN FAO survey[6] found that only about 25 per cent of sawn timber and about 5 per cent of woodfuel was actually taxed.

As long as taxes on urban traders are small they are likely to have little if any effect on curbing forest use. Even if one assumes that everyone pays them, the taxes can easily be passed on to the customer and will make only a small difference to final prices. Equally important, woodfuel harvesters or charcoalers are unlikely to change their ways because the truckers have to pay a tax. If anything, such taxes will tend to lower the price which dealers are prepared to pay to the producers, who may well respond by cutting wood more intensively (and destructively) to maintain their income; for instance, by clear felling rather than only partially "cutting over" a stand of trees. If, on the other hand, the taxes are large, there will be even stronger incentives than now for the transporters and traders to evade them.

Some proponents of increased forest taxes and policing argue that none of these problems would apply if forest services were better managed, with more staff and financing, and that this could be achieved with larger revenues from forest taxes (or external donor assistance). Most African foresters lack mobility and are poorly paid. They are open to small bribes by forest users who want to avoid tax payments, and have little incentive to collect taxes for the "bosses" from generally poor forest users. Give them wheels and better management and pay – perhaps by allowing them a share of the taxes they collect – and the whole system of forest protection and revenue raising would run more effectively. This may well be so, but such systems have not been tried and one can only wait and see if they can be made to work.

The third main reason for taxing forest users has a clearly stated aim.

It is to bring the price of open-access wood resources "at the forest gate" up to their replacement value from formal state-owned plantations, or to the level where rural people could compete by producing woodfuels for sale into urban markets.

It is difficult to comment on the feasibility of this approach since it depends so much on local circumstances and the price structure of woodfuel markets. However, a few general points and questions can be raised.

Private wood producers or state plantations would almost certainly benefit by the taxing of open access resources, but only if: (1) the tax was large enough to bring the price for these resources close to that of the small producers or plantations; (2) tax collection was efficient enough to mean that the average tax was also large; and (3) non-forest producers were close to the forests or already had good access to markets. More remote producers with poor market access are unlikely to benefit much, since commercial traders will face higher than average costs in reaching them and would have to offer less than the going price (or squeeze their own margins). In short, higher prices in place A will not necessarily benefit remote producers in place B. Furthermore, as we saw in the Introduction, in many places much of the commercially grown firewood or charcoal comes as a by-product of tree growing for other purposes, such as black wattle grown for tannin in Kenya. Taxing "free" wood from the forest will not help these producers. They may, of course, benefit from subsidies which would make it economic to produce in areas that are now sub-economic, but that is another matter entirely.

The biggest problem, however, is that with few exceptions forest taxes would have to be raised so much to bring them into line with the costs of sustainable production from plantations – or, to a lesser extent, from smallholdings – that the taxation strategy could not be enforced or could greatly harm urban consumers.

A classic example of this point is Malawi, where the government is now considering a proposal by the World Bank that royalties on wood from state forests should be raised by a factor of three to four by the mid 1990s in order to bring them into line with the costs of producing wood in government plantations. At the same time, the number of forest guards would be greatly increased to ensure that royalties are collected. Some details of this scheme and its possible effects on woodfuel prices for urban consumers are presented in Urban Case 5.

However, in contrast to this pessimistic view, in one case firm measures to control forest exploitation and the transport of woodfuels from banned forest areas has worked by forcing a massive shift towards a more sustainable production system. This happened in 1987 in Rwanda after the government prohibited charcoal production in publicly-owned savanna woodlands to the east of the country and imposed tough fines and strict controls on transporters. Some communities imposed further controls and fines and in one case even closed the local charcoal market. As described in Urban Case 6, the proportion of charcoal coming from these state woodlands fell from 90 per cent to 15 per cent, while the difference was made up by a large increase in production from private and community woodlots in the south and south-west. It will be interesting to see if this reversal can be maintained and, indeed, replicated elsewhere.

MANAGING THE FORESTS

The management of natural forests for greater productivity has often been neglected in favour of "industrial" plantation schemes. But interest in natural forest management is now quickening since it needs much lower investment (for example, no land clearing or seedlings), is well adapted to local ecological conditions, and can produce a range of products.

It is not easy to make meaningful cost comparisons because of differences in agroclimatic conditions, costing methods, species and management regimes. However, one such comparison for reafforestation costs in Niger in 1986 gives such a clear advantage to natural forest management that it is hard to believe that it does not have a wide validity. Measured in terms of CFA francs per kilogram of wood produced, irrigated plantations score 170, rain-fed tree plantations 85–130, rural private woodlots 30–60, and management of natural forests only 15–30.[5]

Many experiments are now under way to improve on the rather limited technical (and social) "packages" for improved forest management. This section looks briefly at five technical approaches which appear to be gaining in favour. They all involve either a considerable change in attitudes by forest services or the adoption of fairly novel techniques.

Encouraging Forest "Mining"

Since the "mining" of woodlands is to a large extent an inevitable stage of development, one proposal is that forest departments should encourage the one-off "mining" of presently protected areas of wood surplus as a transitional strategy before other and more sustainable alternatives become available.[7]

Recent remote-sensing studies of woody biomass resources in the SADCC region have shown that in eight of its nine member states the total annual yield of woody vegetation (or in forestry terms, Mean Annual Increment) exceeds the total consumption of wood for fuel.[7] The problem, of course, is that much of this surplus is located in areas remote from the major demand centres, so that transport costs make it less economic than resources that are currently being exploited; or else it is in protected forest reserves and game parks.

However, there may well be opportunities in some places for using these often large resources. Planners in Botswana, for example, are considering one-off "mining" of the massive wood resources in the north and centre of the country, which are within reasonable reach of road and rail, to reduce pressures on the remaining woodlands in the highly populated south-east region, which includes the capital Gaborone. In Zimbabwe, mining of the Mtao forest reserve near Masvingo has also been considered as a stop-gap solution of this kind.[7] In Malawi, recent studies have shown that it would be economic to exploit the large pine plantations around Viphya in the north of the country to make charcoal, and to truck this 280 kilometres to the capital, Lilongwe, and 630 kilometres to the largest city, Blantyre, in the south of the country.[8] Thinnings from pulpwood and timber plantations at Viphya could produce 930,000 solid cubic metres of wood annually for at least ten years, while low-grade wood could provide a further 370,000 cubic metres – enough, in total, to produce 100,000 tonnes of charcoal a year. Probably many other cases can be found where surpluses could be used at reasonable cost and lower final prices than other proposed alternatives.

Improved Extraction Methods

Many conventional forestry operations are designed with the principal product – such as sawn timber or construction logs – in mind. Alternative management strategies such as selective rather than clear felling, or a mix of felling and pollarding of main branches, can greatly increase sustainable outputs as well as the proportion of saleable woodfuel to principal product, while at the same time reducing environmental risks. African foresters have often learned their craft from northern models which may well be inappropriate to the very different resources, needs and economics of their own countries (see Figures 1.1 and 1.2, p.34, 43, for example). One crucial point in this context is that when they are stressed by pollarding or branch lopping, etc., some tree species greatly increase their growth rate. Preliminary research in the Sahel has recently suggested that such productivity increases may be as large as a factor of ten.[9]

In a number of places useful volumes of "lops and tops" and other wastes from forestry operations, or wood wastes from sawmills and pulpmills, could also add significantly to urban woodfuel supplies. The SADCC Woodfuel Development project[7] cites some examples. In Zimbabwe the residues left after the tanning of wattle could produce an annual 30,000 tonnes of charcoal. In northern Botswana, commercial exploitation of the Thobe indigenous woodlands produces around 100,000 cubic metres of sawmill residues and almost 9,000 cubic metres of in-forest wastes each year. The idea of transporting it as wood or charcoal to the urban centres in the south-east has been considered but is at present constrained by the high transport costs over distances of 600–700 kilometres. In Swaziland, high transport costs inhibit the use of large volumes of wastes from the Usutu plantations which produce some 80,000 tonnes of sawn wood and 180,000 tonnes of pulp annually. Possibly in these and other cases, lower transport costs – for example, through the introduction of "back hauling" trips (see Chapter 6) – or higher urban fuel prices will make the use of such resources economic.

Rotating Extraction Areas

This idea involves controlling where woodfuel cutters can operate so that areas which have been cut are given time to regenerate naturally. To some extent, "illicit" woodfuel cutters already practise this technique, since they often leave some standing timber when they move on to fresh areas. If forest services tried to encourage this tradition rather than trying to restrict "all" unregulated cutting, the productivity of natural woodlands could be greatly enhanced while environmental damage would be reduced.

Enrichment planting

This well-recognized technique involves "in-filling" thin or environmentally vulnerable areas of natural forest with plantings of either indigenous or faster-growing exotic species. Although the technique is practised in Africa its use could be increased in many places if forest departments switched their approach from one of "protect and conserve" towards a more positive attitude of forest utilization and management. A variant of this technique which has not yet caught on so widely and also demands new attitudes by conventional foresters is the planting of buffer zones of fast-growing exotics around natural forests. Local people and commercial woodfuel cutters are allowed to use the buffer zones (under licence), thus reducing the chances that they will also take wood from the protected forest area.

Fire Protection

After clearances for agriculture, bush and forest fires – almost invariably caused by people – are probably the greatest hazard to natural forests, especially in the less arid zones of Africa where grass growth between trees is usually intense.[10] Reasons for starting fires include land clearance for cultivation, hunting, to obtain new shoots of green grass for grazing, to keep paths open, and "revenge against neighbours or government".[10] There is often an attitude that there have always been forest fires – so why worry about them?

The enormous difference to forest productivity that can be made both by fire protection measures, such as firebreaks, and by early-rather than late-season burning for grass control are illustrated in Table 7.1. The difference between the worst and best practices is more than sixfold in terms of the numbers of growing stems and sevenfold for larger growing stems of over 50 centimetres diameter. In the long term the most effective way of gaining these huge production increases is to change attitudes by demonstrating that fires are not inevitable, do a great deal of damage, and can be prevented. This requires a good rapport between foresters and local people, which will be difficult to foster so long as forest services take a restrictive "conserve and protect" attitude to forest management.

Table 7.1: Effects of Fire Treatments on Coppice Growth in
**   *Combretum* Forest in Nigeria**

	fire protected	early burn & fire protected		early burn every year	late burn every year
		burn every 4 years	burn every 2 years		
Stems per hectare	6,664	4,762	3,627	3,639	1,038
Stems over 50 cm girth per hectare	119	85	49	42	17

Note:
Data are for 22 years after start of field experiments.

Source: Jackson *et al.* (1983),[10] from Onochie (1964).[11]

COMMUNITY CONTROL OF THE FORESTS

If the policing and guarding of natural forests by state agencies is proving to be so ineffective, why not give the tasks of control and management – and the revenues received – back to the local communities to whom the forests once belonged? This radical turn-round is now widely accepted as an exciting but difficult way ahead towards both a more productive and a more sustainable use of forest resources on public lands.

The Guesselbodi experiment in Niger is one instance where this approach has been tried and appears to be working very successfully in terms of both greatly increased natural resource productivity and income for the local community: see Rural Case 19. Another advantage of this approach is that it reduces the number of forest department staff and administrative burdens needed to manage the forests, while local foresters, rather than being policemen, have to adopt the more positive role of extension agents. However, no one should think that introducing this type of management will be simple or easily replicated on a standard pattern. Merely handing over natural forests to the local community, as their own property and without conditions, could lead to a lack of all management, or even destruction of the forest, if the community tried to make large but once-and-for-all returns from its new acquisition.

The kind of approach that needs to be taken and the conditions that should be set to ensure sustainable management practices depend a great deal on forestry officials' present control practices and relationships to the local community, legal structures regarding forestry and land use, and all the issues of indigenous knowledge, tenure, community cohesion and the like, which were discussed in Part I. However, the issues differ somewhat because management to produce saleable outputs and cash returns – rather than just the support of rural livelihoods – is basic to the aims of control transfer. Thus, for example, there must be markets for the produce from the forest (such as wood, fuelwood, charcoal, hay and grazing) and there must be robust community mechanisms for sharing benefits and paying for labour. These points imply a great diversity of issues and approaches to them: there are no blueprints.

In particular, where forests belonged to or were freely used by the local community within living memory, but forestry officials have adopted a strong policing role, there may well be a good deal of resentment or apathy towards schemes for community forest management. As Thomson has commented:[12]

> Rural people will not, however, automatically move to manage renewable resources, nor will they be moved by propaganda urging conservation consciousness and participation in sustained-yield management unless (1) they see a real need for such investments of time, energy and money, and (2) technical, economic, financial, legal and political constraints they currently face are overcome or reduced to the point where participation becomes feasible.

A third proviso is that "local people are convinced they will gain real advantages, within a relatively short period, as a result of their efforts. These advantages should preferably be financial ones."[10]

In conclusion, while it may seem unnecessary to raise yet again the question of ensuring full participation, there have been cases of so-called participatory forest management schemes which ended in disaster because the initiators lacked the patience and sympathy fully to take into account the villagers' point of view. The experience of one such project in the Casamance region of Senegal has been summarized concisely as follows:[13]

Start of the programme.........Mistrust......................Refusal to co-operate
Work begins ....Hope ....Massive co-operation
Increased activity ....Fear of deception.......New demands on villagers
Back to status quo ....Hostility ....Sabotage (burning the forest).

PERI-URBAN PLANTATIONS

The idea that urban woodfuels should be supplied from large-scale dedicated plantations close to the cities seems eminently sensible. It also has many technical and bureaucratic advantages. The costs and benefits of peri-urban plantations are relatively easy to define. Land requirements, personnel needs, the timing and volume of investments, recurrent costs and sales can be defined by conventional skills. Well-known technical packages can be compared and the most economic choices applied, regarding, for example, the choice of species, tree spacing, irrigation versus rain-fed, use of fertilizers, and so on. The skills to apply these techniques are readily available. And, not least, a few relatively large plantations are easier to administer than many small-holder enterprises.

There have also been several successes with peri-urban plantations. The most famous examples are the 20,000 hectares of *Eucalyptus globulus* which stand around Addis Ababa and the further 50,000 hectares, mainly of eucalypts, around other Ethiopian urban centres. Although most of these areas were planted within the past twenty years, some plantations date back to the start of the century. Furthermore, their productivity can be remarkably high: a survey in 1980 found that per hectare average yields were 15–20 cubic metres a year, with standing stocks of 118 cubic metres, implying an acceptable five- to eight-year rotation and a large output of some 1.4 million cubic metres of wood a year.[14]

Despite all this, the track record of formal, state-run fuelwood plantations has not been good, especially in dry areas. In such areas plantations are typically very expensive, with establishment costs of around US$800–1,200 per hectare, but occasionally reaching as much as US$7,000–9,000 per hectare for irrigated plantations in difficult conditions in the Sahel.[15] Outputs have often been disappointing. In wetter areas, there has been greater success with lower costs. For instance, estimated establishment costs in Kenya and Tanzania for forest department plantations are shown in Tables 7.2 and 7.3 to be around US$160–280 per hectare, although the wood they produce is

Table 7.2: Comparison of Woodfuel Production Methods in Forest Lands, Kenya

	Model 1	Model 2	Model 3	Model 4	Model 5
Tree management	*forest dept.*	*forest dept.*	*small-holders, forest workers*	*private sector improved*	*private sector improved*
Trees per hectare	1,920	3,000	3,000	3,000	5,332
Establishment costs					
shillings per hectare	2,716	4,043	1,171	1,246	2,716
US$ per hectare	160	238	69	73	160
Wood production costs					
shillings per m^3	99	78	41	36	33
US$ per m^3	5.8	4.6	2.4	2.1	1.9
Charcoal production costs					
shillings per 30 kg sack	54	39	27	39	39
Discounted returns to tree growing & charcoal making	−22%	−12%	31%	8%	9%

Notes: Model 1 has traditional charcoal kilns; with improved kilns, charcoal costs 39 shillings a sack and the annual discounted return is −8%. Costs and benefits are discounted at 15 per cent annually for small-holders and the private sector, at 10 per cent for the forest department.

Source: UNDP/World Bank (1987).[16]

very expensive compared to smallholder production and present taxes on wood from natural forests. There are also positive examples of plantations around Eldoret and Nairobi (Kenya), Mbeya (Tanzania), and Blantyre (Malawi).[2]

However, a general problem with such plantations is that when sited close to urban markets to cut transport costs, they can become increasingly uneconomic as cities expand and land values, wage rates and the profitability of alternative productive uses of land increase in their vicinity. Populations who are displaced by the plantations have sometimes been integrated as part of the forestry labour force, but have more often been evicted to become part of the mass of urban dwellers.[7]

As with the changing attitudes to natural forest management, the new paradigm for peri-urban plantations is to adopt less formal, "softer" methods which involves mixtures of agriculture, agroforestry and woodlots planted and run by smallholders, or much lower-cost techniques of formal plantation management. The economic advantages of these new methods are so great that, once people began (only

Table 7.3: Comparison of Woodfuel Production Methods, Tanzania

(data are per hectare)	peri-urban plantation	smallholder woodlots	forest management and enrichment planting
Seedlings	2,500	2,000	–
Rotation period (years)	26	26	20
Annual wood produced (m³)	15.2	15.2	2*
Establishment costs:			
shillings	24,934	6,695	2,359
US$	277	74	15
(of which for mechanization)	(44%)	(3%)	(4%)
Discounted costs (shillings)	31,036	6,336	2,499
Discounted production (m³)	99	67	27
(discount rate used)	(10%)	(15%)	(10%)
Break-even stumpage price			
shillings per m³	315	95	93
US$ per m³	3.5	1.1	1.0

Notes:
* 21 m³ are produced in Year 1 from thinning and firebreak clearance; production thereafter is 40 m³ for the 20-year rotation. Discounted costs and production are for the full rotation.
m³ = (solid) cubic metre. US$ at official exchange rate.

Source: UNDP/World Bank (1987).[17]

quite recently) to compare the economics across the whole spectrum of "high" versus "low" technology approaches, the choice of the latter becomes almost inevitable.

The chapter concludes with two such recent comparisons for Kenya and Tanzania. Although it must be noted that these costings are no more than pre-project exercises – and that actual outcomes might be very different – it is not hard to see why the shift away from "industrial" to smallholder models is now under way.

For Kenya, the UNDP/World Bank's Urban Woodfuel Development Programme found that large, formal plantations under direct Forest Department management "appeared not to be viable"[16]. Instead, it made the estimates shown in Table 7.2 (p. 233) for five types of sustainable wood growing for charcoal production on gazetted forest land, thus allowing the forest establishment costs to be offset by sales of wood from the initial forest clearing. Only the smallholder and "improved private sector" models were economically viable, largely because their costs for wood production were as little as 33–41 per cent of those for the Forest Department methods.

For Tanzania, the UNDP/World Bank's Urban Woodfuels Supply Study[17] compared: (1) a conventional peri-urban plantation run by the Forest Department with (2) smallholder woodlots in a *taungya* system (in which food and tree crops are produced in forest land), and (3) management of the natural forest with some enrichment planting. Summarized findings are shown in Table 7.3 (p.234). Although costs and benefits for the smallholders are discounted at 15 per cent a year, while the other (government-run) options have only a 10 per cent discount rate, the smallholder model is clearly by far the cheapest even though it produces as much wood as the formal plantation; that is, just over 15 cubic metres per hectare a year. Establishment costs are only one-fifth as much, while the proportion of this for mechanization with its high foreign exchange component is only 3 per cent (for seedling transport) compared with 44 per cent for the much higher total costs of the formal plantation. In fact 90 per cent of the costs for establishing the smallholder woodlots are for seedlings: probably cheaper, decentralized methods of seedling production could be used.

The final row of Table 7.3 shows the sale price for wood that is needed to break even: that is, the discounted production in terms of wood volume divided by the discounted costs over the rotation cycle. For smallholders, this price is 95 shillings (or just over US$1 at the official exchange rate) per cubic metre – or under one-third that of the plantation (315 shillings or US$3.5 per cubic metre). However, the smallholder model includes the costs of food production and harvesting but in the economic sums counts only the wood production as a benefit. Natural forest management with enrichment planting has even lower establishment and discounted costs and its required stumpage price is very slightly less than for the smallholders. However (on these estimates), wood production is only 2 cubic metres a year after the initial clearances, or one-seventh as much as in the other approaches.

Since the present stumpage royalty for collecting wood from natural forests is (if paid at all) only 10 Tanzanian shillings per cubic metre, none of these approaches can at present compete with "free" wood from the forest. But since the sustainability of this supply source is in question, there seems to be a clear case for further studies and pilot schemes aimed at "proving" the two low-cost options and especially the small-holder woodlot model and variants of it. As discussed in Part I, there may well be other smallholder wood production models which are equally productive and more acceptable to the smallholders themselves.

As for the plantations, the UNDP/World Bank study suggests that they will never be economically viable for firewood or charcoal markets but may find a place for producing higher-priced construction poles. There may also be other extremely good reasons for implementing such schemes, including employment and income generation and environmental protection. These are all important justifications for afforestation. Indeed, as we have seen in many other places in this book, woodfuel supply itself can rarely be the primary justification for beneficial change. It is usually best seen as an indirect benefit from broader actions and policies.

Chapter 8

Fuel Switching and Saving

A major shift from the use of traditional biomass fuels to "modern" fossil fuels and electricity has always been a fundamental feature of economic growth. Whereas woodfuels account for 60–95 per cent of total energy use in the poorest developing countries, in middle-income countries they account for only 25–60 per cent, and in high-income industrialized countries, with minor exceptions, for less than 5 per cent.[1,2]

On the national scale, these so-called "energy transitions" are partly due to the extra-rapid growth in the normal development process of industries and forms of transport which use modern fuels. But they are also driven by the mass of households and small commercial consumers who link the advantages of modern energy sources with aspirations to higher living standards in the home and greater productivity in the workplace. As incomes rise and more people and businesses can afford to pay for the greater convenience, cleanliness and versatility of modern fuels, the energy system is forced to respond by trying to improve their supply and distribution.

The main question addressed in this chapter is whether attempts to force such a shift provide a feasible and useful answer to the woodfuel problem. Can woodfuel-modern fuel substitutions be speeded up at the consumer level, and by what means? Why should one try to do this? To increase consumer choice and security of supply? To "save the forests" and reduce the need for afforestation to provide adequate woodfuel supplies? And what are the wider economic implications, such as the impact on debt and import bills?

We also consider, though briefly, the main alternative to this approach for reducing woodfuel consumption: the promotion of more efficient woodfuel cooking stoves. Although it may seem odd to mix these seemingly different strategies together, we do so because from the consumer's or a government's viewpoint they are closely connected aspects of demand management and, to some extent, are in conflict.

In dealing with these questions we shall come up against the same generic problems that we have met previously: the lack of sound data combined with the enormous diversity of local conditions. Consequently, the chapter concentrates on the underlying issues of the transition which can be generalized and uses specific examples only as illustrations of each point. Although some of these illustrations do have a broad validity, effective policies and actions must of course be built on the specific circumstances of each place.

IS THE ENERGY TRANSITION HAPPENING?

The first question that must be asked is whether the energy transition is occurring in Africa. The frank answer is that we do not know, except for rare instances where circumstances are highly exceptional. For instance, it is known that in Addis Ababa (Ethiopia) households have been switching very rapidly from firewood to kerosene because firewood prices have been rising steeply.[3]

More generally, the question cannot be answered because governments have failed to collect the vital data. All countries know how their use of modern fuels is moving over time. Many can make reasonable estimates of the consumption of these fuels by households and other large woodfuel-consuming sectors. What is not known is how woodfuel consumption is changing over time, and what the resulting degree of substitution is. For instance, in the mid-1980s price changes in Niger for bottled gas (LPG, or liquefied petroleum gas) led to a 60 per cent rise in consumption within three months.[4] However, for lack of information on past and present household energy use no one knows: (1) how much of this change was due to greater consumption by existing LPG users; (2) how much was due to new LPG users and what fuel they switched from; and (3) the effects of all this on woodfuel consumption.

Where there is the information to answer the question, it suggests that the household transition can occur very rapidly, in low-income countries, if the conditions are right. In India, large urban household surveys were conducted in 1978–9 and 1983–4 by the same organization, so that comparisons between the two periods are probably fairly reliable.[5,6] Firewood prices rose during this time compared to all the competitive modern fuels – kerosene, LPG and electricity – while prices of the latter two fell in real terms. More importantly, incomes rose and were more evenly distributed, so that the number of middle-income families increased sharply. As a result, in the average Indian city the share of energy for cooking and heating which was met by firewood fell in only five years from 42 to 27 per cent, while the shares for kerosene and LPG nearly doubled from 19 to 36 per cent and from 7 to 12 per cent respectively.

The lack in most African countries of such basic information on energy demand trends is itself a major policy issue which needs urgent attention.

However, we can reasonably ask what are the "right" conditions for fuel switching on this scale to occur. Or, in other words, what drives the energy transition?

It is now clear from many surveys – particularly of household energy patterns – that there are two main driving forces: (1) access to dependable supplies of modern fuels in sufficient quantity, and (2) sufficient income to invest in the devices for using them or to overcome other barriers to adopting modern fuels. These are interlinked. Fuel prices appear to play a much more minor role but can help to speed substitutions "forward" into modern fuels or start a "backward" shift from modern fuels into woodfuels.

FUEL SWITCHING AND URBAN SIZE

The effects of fuel access can be seen most clearly in the patterns of household energy use according to settlement size. In remote rural areas, even high-income families typically depend almost entirely on biofuels, with perhaps a little kerosene for lighting or kindling the cooking fire. Supplies of modern fuels are either non-existent; or they are insufficient, unreliable or very expensive for the major end-uses which biofuels satisfy: cooking and water and space heating.

For example, during the early 1980s in average rural conditions in Brazil, India, Pakistan and Sri Lanka across a large income range there was virtually no reduction in the 90–95 per cent dependency of households on biofuels.[2,5] This must have been due to problems of fuel availability rather than cost, since incomes in the upper range were as much as US$18,000 a year (1975 rates corrected for purchasing power parity). The same must be true of Africa, though there is little information on rural energy use by income to corrobate this.

As a country develops and/or one moves towards the larger towns and cities, access to modern fuels increases, since these areas are the centres of demand and economic power and have the best-developed infrastructure for distributing modern energy sources. This "gradient of opportunity" has been better mapped for Asia than Africa but the implications must be similar.

For instance, during 1979 average Indian towns with a population of 20,000–50,000 met just under 40 per cent of residential energy use with modern fuels. In the 50,000–200,000 population range the proportion jumped to 56–58 per cent, with a steady increase to 66 per cent in cities of 200,000–500,000 inhabitants and 75 per cent in even larger cities.[5,6] Even more strikingly, a household census in Pakistan[7] found that dependency on woodfuels for cooking and heating was 98 per cent in rural areas, 93 per cent in small towns of under 25,000 people, and around 65–75 per cent in medium-sized cities; the proportion dropped suddenly to only 25 per cent in the largest cities of over 200,000 inhabitants. Furthermore, these largest cities accounted for roughly three-quarters of the urban population and its growth, thus accentuating the rural-urban differences.

Other things being equal, the energy transition is therefore driven by urbanization and, in particular, the size and growth of the largest "primary" cities. History, geography and politics also play their part, of course. The location of fossil fuel and hydro-power resources, policies on rural and small-town development, the costs of road transport and the extent of road networks, and many other factors, can all affect the availability of modern energy sources outside the large cities.

The big questions in this context for Africa are: (1) how rapidly will the large primary cities continue to grow? (2) can their modern fuel supply and distribution systems keep pace with growing demand? and (3) how quickly will the fuel access advantages of the large cities spread outwards to the countryside where the vast majority of potential modern fuel consumers live?

FUEL SWITCHING AND INCOME

Across this broad rural-urban gradient of opportunity lies a second income-led transition. More accurately, within each city and its particular if changing access to modern forms of energy there is a gradient in the use of modern fuels which is driven by fuel preferences but constrained by various cost barriers. Higher income allows one to satisfy the preferences by jumping the barriers.

A classic income progression of this type is shown in Table 8.1, for

**Table 8.1: Residential Energy Use by Household Income: Urban
          Kenya, 1980**

income group (thousand K sh.)*	per cent of total for each income group					total GJ/year†
	firewood	charcoal	kerosene	LPG	electricity	
0 – 3.1	63.3	28.2	8.5	–	–	31.6
3.1– 9.1	25.5	57.4	17.1	–	–	33.3
9.1–18.2	16.4	64.2	16.5	1.6	1.3	37.1
18.2–54.6	6.4	61.3	14.4	9.4	8.5	36.2
above 54.6	1.7	32.0	4.1	14.6	47.6	41.8

Notes:
* Kenya shillings. US$1 (1980) = 7.42 K sh.
† Conversions: firewood 16 MJ/kg, charcoal 30 MJ/kg, kerosene 35 MJ/litre, LPG (liquefied petroleum gas) 45.5 MJ/kg, electricity 3.6 MJ/kWh.

Source: Adapted from O'Keefe et al. (1984).[8]

average towns in Kenya in 1980. Firewood is the dominant fuel for the poor, while firewood and charcoal use combined decline with higher income. Kerosene is used by the poor, mainly for lighting. Its use increases with income but then declines sharply as it is displaced by more desirable bottled gas (LPG) and electricity among the upper-income groups. Much of the electricity use by the rich is for lighting, refrigeration and other modern appliances and does not displace biomass fuels.

Figure 8.1 attempts to put together all of the major issues which underlie this income-led transition. It acts as a key to the discussion of them in the following sections.

Fuel and Equipment Preferences

The underlying reasons for moving up the fuel "ladder" shown in Figure 8.1 are well recognized: households (and other consumers) would like to use energy sources with the widely desired characteristics shown at the top. Most surveys in developing countries show that electricity (and natural gas) are widely considered to be "ideal" for cooking, for much the same reasons as they are preferred and used in rich countries. They are delivered to the home, do not have to be stored, are clean to use, and are versatile and highly controllable. For cooking in particular, they can be switched on and off at will and set to any of a wide range of power outputs for fast or slow cooking.

Bottled gas comes a close second, since it meets nearly all of these criteria provided that bottles are delivered to customers; if not, fetching

Figure 8.1: Urban Fuel Preferences and Constraints

"IDEAL" GOALS	Clean to use. Delivered to user. No storage. Versatile: e.g. good control of heat output. High efficiency holds down costs.

Fuel preference "ladder"		Barriers to climbing the ladder		
		Equipment Costs	Fuel payments	Access
ELECTRICITY		Very high	Lumpy	Restricted
BOTTLED GAS (NATURAL GAS)		High	Lumpy	Often restricted, bulky to transport
KEROSENE		Medium–high	Small	Often restricted in low-income areas
CHARCOAL (may be higher in some cultures)		Medium	Small	Good: dispersed markets and reliable supplies
FIREWOOD / CROP RESIDUES ANIMAL WASTES	Possible conversion to high-grade energy forms (e.g. biogas, electricity)	Low/zero	Small or zero	Good: dispersed markets and reliable supplies. Can usually be gathered "free"

them in all but the smallest sizes demands wheeled transport – a major constraint for the poor. At the other extreme, firewood (and the even lower-grade fuels shown below it) scores badly on all these counts. It is bulky to take home, smoky in use (although this depends on the cooking equipment) and hard to control. Storage may be a critical difficulty in the small houses or single rooms of the poor, where dampness and termite attack can also pose serious storage problems. Charcoal and kerosene are intermediate, while choice is affected by cooking method and taste preferences. In many East African cities, for example, charcoal is often used by all income levels for grilling and other cooking tasks requiring high power outputs, not least because of the taste imparted to the food by the charcoal smoke.

As these remarks hint, the devices for using fuels are as important a consideration as the fuels themselves – a fact which is sometimes ignored in studies or policies on fuel switching. A good illustration of this point[4] comes from eastern Niger, where it is much cheaper to cook with kerosene than with wood. However, kerosene is hardly used as a cooking fuel, since most of the available kerosene stoves suffer from three defects as far as local consumers are concerned. Their maximum power output is only 2 kilowatts (kW), compared with the 6 kW or more for the typical wood fire, so cooking by kerosene takes much too long. Second, kerosene stoves are not designed to take the spherical cooking pots which are used in the area: pots can easily tip over with the frequent stirring common to local cooking practices. Third, the kerosene stoves are not very sturdy and are liable to break.

For reasons of this kind, kerosene is very frequently used only for rapid cooking and water-heating tasks such as making hot drinks, while firewood or charcoal is used for the main staple dishes which take longer to cook or need high heat outputs, as for grilling and frying. For reasons of cost LPG and electricity are also frequently used in the same way as kerosene by all but the richest households. A recent household energy survey in Dar es Salaam, for instance (see Urban Case 4), found that electric cookers were mainly used only for making breakfast and hot drinks in the evenings.

Housing conditions can alo have major effects on fuel and equipment choice. The poorest dwellings of the shanty towns are normally too flimsy and unsafe to support electric wiring, and such areas may not have an electricity supply at all. Kerosene and bottled gas equipment are valuable items which may be stolen and are too bulky to move easily. Many shanty town families live by migrant labour and can be evicted overnight. Higher up the income and employment ladder, the apartment blocks of industrial workers and the like frequently have no flue or chimney, while cooking outdoors is at best difficult: smoky woodfuels are virtually ruled out.

Access to Fuels

These ideal preferences are modified or constrained by several factors, of which access to fuels and the reliability of supplies are among the most crucial. Access problems are major reasons for fuel switching as well as barriers to it. They also explain why multiple fuel use for such essential services as cooking is the norm in most urban households as it minimizes risks of supply failure.

Furthermore, access and reliability problems can be peculiarly local. With kerosene, for example, there may be no incentive for retailers in one poor district to stock more than small amounts for lighting. In another district more enterprising shopkeepers and/or different cooking habits may have encouraged a substantial market and steady demand. Bottled gas supplies are notoriously erratic but, as with kerosene, available supplies are often routed to higher-income districts where demand is both larger and more dependable.

The impact of these local factors on fuel use and fuel switching should never be underestimated. For the poor, access issues frequently outweigh all other considerations, including equipment costs and fuel prices. Turning to Asia again, a 1985 survey in Lucknow found that few poor families cooked with kerosene even though they realized that it would cost only 40 per cent as much as cooking with firewood.[9] With average daily incomes of around US$1.0, the cost of a kerosene stove (US$1–4) was not seen as a major deterrent. Kerosene shortages and long queues at the shops were given as the main reason for not using the fuel. Similar constraints are common in other cities in India[10] and in Africa, where cooking stoves and lamps for using modern fuels, as well as the fuels themselves, are often hard to obtain. This has certainly been the case in war-torn Angola and Mozambique, for example.[11]

Equipment Costs

Figure 8.1 (p.242) also shows that a second crucial factor in fuel switching is the cost of fuel-using equipment. Recent data from Kenya illustrate how great these investment barriers can be. A 1986 survey by the FAO[12] found that compared with a zero-cost traditional three-stone fire, a standard charcoal stove cost 20–50 shillings and the famous improved charcoal *jiko* stove cost around 60–100 shillings (US$4–7), depending on its size. According to interviews with the *jiko* disseminators in September 1987, even the latter price was considered too high by most of the target group of poor families, who nevertheless appreciated that *jikos* would give substantial fuel and cost savings.[13] Kerosene stoves (mostly imported) cost much more at about 450 shillings (US$30). However, switching to bottled gas required an investment of at least 3,100 shillings or US$206: 1,440 shillings for the deposit on a 15-kilogram gas cylinder and 1,650 shillings for a two-ring stove. A four-ring stove with an oven providing the full range of cooking facilities cost 4,350 shillings or more.

In other cities, equipment cost differences may be much smaller. A survey of stove prices in Gaborone, Botswana, in September 1983 found that a metal woodfuel stove (*louga*) cost Pula 10 (US$9), or much the same as the Pula 11.5 for a non-pressure kerosene stove.[14] A one-ring gas cooker cost only three times as much (Pula 28–30), while kerosene pressure stoves were priced as high as Pula 57. Clearly, there are no simple rules about device costs or their affordability. Much depends on the abilities and raw-material costs of local manufacturers, and presumably also on levels of demand as they affect economies of scale for production and marketing.

Another important investment constraint for the mass of poor urban consumers is that electricity and bottled gas must be paid for in large lump sums. Woodfuels and kerosene, on the other hand, can be purchased in small amounts for a little cash on a daily basis. Avoiding "lumpy" payments is a major household fuel strategy for the poor, even though it often means paying up to twice as much per unit of firewood, charcoal or kerosene as for larger volumes. In part this is due to the very real additional retailing costs for "bulk breaking" to smaller sizes – itself a vital service provided by the fuel traders – but also because traders sometimes take advantage of this consumer problem and charge excessively high prices.

Permanent Versus Flexible Fuel Switching

All these factors of fuel and equipment choice, together with fuel prices, imply that fuel switching is driven by two crucially different processes.

The first process is based on *acquiring energy-using equipment*: for example, when a wood-using family buys a kerosene or gas stove. This is obviously the main process behind large-scale fuel substitutions up the energy preference ladder, typified by the progression shown for Kenya in Table 8.1 (p.241). Since the move into new equipment can involve large investments, especially for bottled gas and electricity which provide the largest range of desired fuel/equipment characteristics, it is strongly income dependent – or poverty constrained. It is also fairly permanent and robust since, once acquired, new equipment is rarely discarded. This process therefore sets the longer-term trend of fuel consumption patterns.

The second process involves *shifts between multiple fuels.* Having acquired a kerosene stove, the example family can use kerosene and wood. How much of each they actually use will depend on complex adjustments between preferences and cooking habits on the one hand, and access, availability and price factors on the other. Although rather little is known in detail about this process, it can obviously be very rapid and flexible. This type of fuel switching can go "backwards" or down the preference ladder, when it can be very rapid since there is no need to buy new equipment. Or it can go forwards up the ladder – but only to the extent that the existing stock of energy-using equipment owned by the family – or city – allows.

Intuition suggests that this latter process can also have substantial long-term effects if there are large and sustained differences in fuel prices and fuel availability. High prices or shortages of modern fuels compared to biofuels, for example, can easily lead to "permanent" backward switching of this kind. This has occurred, for example, in Sri Lanka after the kerosene price rose far above the firewood price.[5] The reverse condition, where modern fuels are cheaper than biofuels (they are rarely more readily available) is likely to have a much lesser impact because only a minority of consumers typically owns modern fuel-using devices.

Many of these points can be illustrated by an actual example of fuel switching behaviour in an African city (see Urban Case 4). In January 1987 a Household Energy Consumption and Cooking Habits Survey was conducted in Dar es Salaam (Tanzania) under the auspices of a World Bank project.[15] Although the sample size was small (230 households) and only 12 per cent of these had switched cooking fuels in the previous two years, the survey produced some very interesting results. One was that there had been more backward than forward fuel switching, even though this generally meant switching into a more expensive fuel. A quarter of the families which had switched fuels moved from charcoal to firewood, probably in order to reduce costs since all but 7 per cent of the firewood users gathered firewood rather than buying it. Just over a third moved from kerosene to charcoal. Another 10 per cent moved "backwards" from bottled gas or electricity to charcoal. These proportions considerably exceeded the numbers moving up the normal preference ladder from firewood to charcoal or from charcoal to bottled gas and electricity.

The second key point was the large variety of reasons given for fuel switching. While cheapness was the largest single reason for switching fuels at all and for choosing the new fuel, taken together many other considerations – mostly to do with supply difficulties and convenience of use – were judged to be more important than fuel prices alone.

Although no definitive conclusions can be drawn from such a small survey in one city, these findings do support intuition and much anecdotal evidence that convenience of use, ease of access to fuels, and reliable supplies, are extremely important considerations to the majority of fuel users. Typically they are more important in total than relative fuel prices – unless, perhaps, these price differences are very large.

FUEL PRICES

Governments find it notoriously difficult to control the prices of firewood and charcoal because they are distributed and sold in dispersed and "informal" markets. But of course they can and do control the prices of modern fuels. They do so for various reasons: to extract revenues from fuel taxes, to attempt to contain demand by raising or maintaining high prices through taxes (or by not applying subsidies when world prices rise), or to hold down high and rising prices with subsidies. The latter are normally applied to protect the poor but usually end up benefiting the rich who are the main users of modern fuels.

If such controls produce large and sustained differences between the prices of woodfuels and modern fuels they will obviously have a significant impact on fuel choice and woodfuel-modern fuel switching. They will do so immediately through short-term switching between multiple fuels. And they can have longer-term and more substantial effects by encouraging demand for modern fuels; this in turn forces the manufacturing and fuel distribution systems to provide adequate fuel-using equipment and improve fuel access.

Unfortunately, though, very little is known about the realities of relative fuel prices in any Third World city. This is not because the prices of the fuels themselves are unknown, but because there is virtually no reliable information about the efficiency with which they are actually used in major tasks such as cooking. The consumer is of course not so much interested in the price of energy as purchased but in its price per unit of "effective" or "useful" energy for a particular task: for instance, the price of "heat under the cooking pot". In other words, efficiency of use is critical.

While it is known that electricity and gas (and to a lesser extent, kerosene) are generally very much more efficient than woodfuels, and while thousands of laboratory and field tests have measured the efficiency of every known fuel/equipment combination from three-stone fires to electric kettles and ranges, real cooks and the way they manage their cooking equipment surround these figures with huge uncertainties. Furthermore, even the efficiency ranges found in laboratory tests for any particular fuel vary widely depending on the type of equipment and cooking utensils employed; for example, whether clay or aluminium pots are used, whether lids are kept on or off, or whether fires and stoves are screened from the wind when cooking is done outside. These variations often exceed the known differences in fuel prices. As a result, comparisons of "effective" or "useful heat" prices of different fuels depend as much as anything on what efficiency one picks on in the possible range.

This crucial point is illustrated in Tables 8.2 and 8.3. Table 8.2 provides a summary of cooking efficiencies for virtually the full range of fuels and cooking appliances, based on laboratory and field tests. The generally superior efficiencies of modern fuels are apparent, but so are the large ranges of efficiency for each fuel and many of the fuel/equipment combinations. For example, the measured efficiency of traditional three-stone open fires is shown to vary nearly fivefold, from 5 per cent to 24 per cent.

Table 8.3 presents the prices of cooking fuels and electricity compared with those of firewood in the principal cities of some African countries in recent years. The prices are per unit of energy "as purchased" rather than the price for useful heat; that is, not allowing for the efficiency of cooking devices. The variations of relative prices across the cities are enormous and, of course, can change rapidly in each city from year to year. In some cases it is quite clear that when cooking

Table 8.2: Cooking Efficiencies for Various Fuel and Equipment Combinations

fuel/equipment type	efficiency (per cent)	
	laboratory tests	field tests
Firewood		
open fire (clay pots)		5–10
open fire (aluminium pots)	18–24	13–15*
Ethiopian ground oven	–	3– 6
mud/clay stoves	11–23	8–14
brick stoves	15–25	13–16
portable metal stoves	25–35	20–30
Charcoal		
mud/clay stoves	20–36	15–25
metal (ceramic lined) stoves	18–30	20 35
Kerosene		
single wick	20 40	20–35
multiple wick	28–32	25–45
pressure stoves	23–65	25–55
Bottled gas (LPG)		
butane	38–65	40–60
Electricity		
single element	55–80	55–75
kettle or "jug/pot"		80–90

Notes:
Efficiencies assume the use of aluminium cooking pots unless otherwise stated.
Laboratory tests are mostly for boiling water; field tests generally reflect the cooking of a standard meal.

* Based on a very limited range of field tests: actual range is certainly much larger.

Source: Leach and Gowen (1987).[2]

efficiencies are allowed for, modern fuels have a large price advantage or disadvantage compared with woodfuels. For example, in Sudan modern fuels are extremely expensive and in Morocco and Tanzania very cheap. The main point, though, is that in many cases the price differences are smaller than the uncertainties about fuel and device efficiencies shown in Table 8.2. In other words, it is easy to make modern fuels appear more or less expensive on a useful heat basis merely by picking an appropriate efficiency from the ranges shown in Table 8.2.

The main conclusion of all this is that laboratory and desk studies

Table 8.3: Relative Prices of Cooking Fuels for Some African Countries

		firewood	charcoal	kerosene	LPG	electricity
Botswana	1983	1.0	3.0	2.6	4.1	6.9
Burkina Faso	1986	1.0	—	2.4	—	—
Ghana	1986	1.0	1.1	3.2	2.1	3.5
Ivory Coast	1986	1.0	2.4	4.4	4.1	—
Kenya	1985	1.0	1.2	3.0	3.9	4.9
Malawi	1985	1.0	2.3	7.0	—	7.2
Morocco	1985	1.0	—	0.9	0.5	1.5
Senegal	1986	—	1.0*	—	4.6	—
Sudan	1986	1.0	2.0	9.6	12.2	6.8
Tanzania	1984	—	1.0*	1.0	—	0.8
Zimbabwe	1986	1.0	—	4.3	4.6	2.5

Notes:
Prices are for the capitals or principal cities, except for Ghana (average cities). They
are per unit of energy content as purchased, not per unit of effective or useful heat.
Energy contents assumed are as in Table 8.1. Actual market prices rather than official
prices are used, but since these vary considerably the data are only approximate.

* Prices set to charcoal = 1.0.

Sources: As Tables 6.2 and 6.3.

which compare fuel prices are of little value on their own. They must be
backed up – or replaced – by asking the people concerned what they
think or observing what they do. Such studies on consumer opinions
and behaviour are rare, so that little is actually known about either the
underlying realities of fuel prices or what happens when they alter with
respect to each other. This ignorance makes it a trifle difficult to use fuel
pricing as a tool for accelerating fuel substitutions.

FUEL SWITCHING VERSUS FUEL SAVING

Poor information is also central to the issue of whether one knows if
energy users are more interested in measures to help them switch fuels
or save energy, for instance with improved cooking stoves. Table 8.4
provides a good starting-point for considering this question and is
based on data for Lusaka (Zambia) in 1986.

Column (1) shows a steep increase in equipment costs as one moves
towards more desired fuels and improved equipment. Column (2) gives
(highly uncertain) estimates of fuel efficiency at each step, while column
(3) presents fuel prices. The remaining columns present the monthly

Table 8.4: Equivalent Costs of Fuel/Equipment Combinations in Lusaka, 1986

equipment & fuel	(1) equipment cost (Zk)	(2) efficiency (%)	(3) fuel price (Zk/kg)	(4) monthly use (kg)	(5) fuel only (Zk)	(6) fuel & equipment (Zk)
					monthly costs	
3-stone fire (firewood)	0	10	0.27	297	80	80
Improved stove (firewood)	?	20	0.27	148	40	40
Mbaula (charcoal)	5	15	0.79	106	84	84.3
Improved stove (charcoal)	25	22	0.79	72	57	57.7
Wick burner (kerosene)	36	37	2.67	31	83	84
Pressure stove (kerosene)	170	50	2.67	23	61	63.8
1-plate stove (electricity)	285	63	0.07*	210*	14.7	18.7
cooker & oven (electricity)	1,990	75	0.07*	176*	12.3	38.8

Notes:
Zk = Zambian kwacha.
* kilowatt hours.
Useful heat "under the pot" equals 475 MJ per month in all cases. Heat values as for Table 8.1. Electricity costs exclude Zk1,520 (subsidized) connection fee and estimated Zk1,000 to 3,000 per dwelling for internal wiring.

Source: Adapted from ETC (1987).[11]

costs of using each fuel/device combination, assuming that a family requires 475 MJ (megajoules) a month of useful heat (heat "in the cooking pot"). Column (4) gives the amount of fuel or power that must be purchased. Columns (5) and (6) respectively show the monthly costs for fuels alone and for fuels plus the price of the equipment amortized over its lifetime.

Clearly, on this cost structure (and with these assumptions on cooking efficiencies) there is a negligible cost difference between using firewood, charcoal and kerosene with "unimproved" devices: all cost around 80 kwacha a month. The big cost saving comes with electric cooking, but the investment costs (especially when connection and wiring charges are included) are prohibitive for all but the highest-income families. Furthermore, apart from electricity, the equipment costs are negligible when they are spread out over the device's lifetime. This simply confirms the point that it is the initial equipment cost which is one of the chief barriers to fuel switching, unless some form of credit financing is available.

However, the most telling points are: (1) that more efficient firewood and charcoal stoves, followed by high-performance and efficient kerosene pressure stoves, give by far the lowest running costs after electricity, but that (2) the improvements are within the margins of uncertainty about fuel efficiencies. They are also within the margins of the efficiency improvements which can be achieved by many households through better fire management and cooking practices alone, without buying an improved stove. In short, the acceptability of efficiency improvements can only be tested by trial and error – by the consumers themselves.

Table 8.4 also forces one to ask a key question to which there are few reliable answers for African cities: do people invest up the fuel ladder for non-economic reasons of fuel preference, fuel availability and the like, or to reduce costs?

If the non-economic reasons dominate, each fuel-using group will be more interested in substitutions up the ideal preference ladder than in more efficient devices or other means of reducing costs. For example, low-income charcoal users might be more receptive to improvements in kerosene or bottled gas distribution or efficient, low-cost stoves for these fuels than in fuel-saving charcoal stoves. Subsidies on kerosene and gas stoves or schemes to allow their first costs to be spread out might also be attractive options.

On the other hand, if costs are the dominant consideration, measures to support fuel saving may be more successful than the encouragement of upward fuel switching, although this will obviously depend to some extent on relative fuel prices.

In fact, in any city motives are likely to be mixed and to vary from one social group to another. At the same time, none of the measures outlined above precludes the others, so that a good mix of interventions and policies is almost certainly required.

ENCOURAGING FUEL SAVING

Programmes to design and disseminate improved cooking stoves have been a major feature of woodfuel strategies in developing countries but have had a somewhat chequered history. Around 1983–5, in particular, donor agencies and governments had become hesitant about pouring still more money into them, following influential reviews published in 1983 by Gerald Foley,[16] who suggested that improved stoves saved little if any fuel, and by Kirk Smith,[17] who said they mostly failed to meet the other major objective of reducing smoke and associated eye and lung diseases. This hiccup, however, seems to have come at just the time when a number of stove programmes were beginning to take off and new designs and dissemination methods were being developed which paid much closer attention than before to the real needs and interests of stove users, manufacturers and marketers.

One of the main lessons that is being applied in this new wave is that improved stoves have to "fit the cook". Consequently, a great diversity of designs may be needed to fill particular user requirements: in China, for instance, over 100 stove models have been produced to match local patterns of fuel use (crop residues of several kinds, firewood, timber, coal and charcoal, and whether fuels are used singly or in combination), and differing family sizes.[18] For similar reasons, the adaptation of stove designs can often be intricate, demanding painstaking attention to detail. For instance, the reasonably successful Niger (urban) metal woodstove has evolved from an earlier local design (the Mai Sauki) which was itself adapted from a metal stove used in Burkina Faso (the Ouga Métallique).[19] Amongst the many adaptations which were made to fit local conditions were: manufacture from scrap rather than new steel and elimination of anti-corrosive paints to reduce costs; use of folded metal and riveting rather than welding to match local artisanal skills; large handles to improve portability; and three more minor design modifications to improve the stove's stability.

Taking the stove user's perspective, a recent review of major stove programmes[20] has summarized the various characteristics that an improved stove may have to have, to varying degrees. It must:

- offer a good match with present cooking patterns (cooking functions, size and space or portability, range of fuels and cooking utensils)

- save time (on fuel collection if this is a burden, and on cleaning pots, cleaning the kitchen due to smoke, etc.)
- be more fuel-efficient
- reduce accidental burns
- reduce smoke, and eye and lung disease
- reduce maintenance and have a longer lifetime, and with rural stoves use available materials for repair and construction
- be aesthetically pleasing, providing prestige without creating resentment by neighbours
- act, in development-orientated programmes, as a vehicle for giving women greater confidence in their ability to improve life for themselves and their children.

Greater fuel efficiency and related time savings are clearly high on this list and they have certainly been achieved by many new stoves. The popular version of Foley's report[21] had on its cover the irreverent question: "How much wood would a woodstove save if a woodstove could save wood?" That question would not be asked today. One of the main lessons of more recent, careful reviews of stove performance is that whereas traditional cooking fires and stoves can be very fuel efficient if they are operated carefully, with high-quality fuel, out of the wind – and appallingly inefficient if not – improved stoves tend to have consistently high efficiencies however they are operated.

Recent tests on twelve improved charcoal stoves, for instance, found that at high power outputs fuel efficiencies of 40–45 per cent (or roughly double that of traditional stoves) were obtained, almost regardless of what utensils were used or how the stoves were used.[22] At low power (for simmering and so on), the efficiency of the ceramic-lined stoves went up by 8 to 17 percentage points to rival that of gas or electricity, but for unlined stoves they fell by 3 to 13 percentage points. The recent review of stove experience cited above[20] found that fuel savings in four large rural woodstove programmes were 16, 25, 62 and 63 per cent (with the lowest figure possibly being due to sampling errors or a decline in construction standards among village stove makers). An average 30 per cent saving was recorded for the improved (urban) charcoal *jiko* promoted by the Nairobi-based Kenya Energy Non-governmental Organization (KENGO). These are large reductions which promise considerable savings in consumer costs and demands on woodfuel resources.

However, a major snag of most improved stoves which greatly affects consumer acceptance is a combination of low maximum power output and a relatively small "turn-down ratio" between high and low power output – the former for quick cooking and bringing the pot to the boil, the latter for simmering, etc. For the twelve charcoal stoves mentioned above, maximum power outputs were no more than 1 to 4 kilowatts. For eleven of them the turn-down ratio was only 2:1 to 3:1. Only the twelfth stove – the ceramic-lined Thai bucket stove – had a reasonably high ratio of 5.7:1, rivalling the cooking versatility of a gas or electric cooker.

But as the list of essential stove properties suggests, fuel saving is not everything. This point is clearly underlined by the large and methodologically consistent review of stove experience which was made by the Foundation for Woodstove Dissemination and reported at the October 1987 international meeting of stove designers and disseminators.[20] The remainder of this brief review summarizes the results of this survey.

Seven leading stove programmes were surveyed but usable results were obtained only from five which, between them, had disseminated at least 390,000 improved stoves. One programme, run by KENGO, was concerned with urban charcoal stoves. The other four involved rural wood stoves: Dian-Desa in Indonesia, CEMAT in Guatemala, NADA in north India, and ASTRA in south India. Table 8.5 summarizes some of the findings.

In all five programmes, the majority of people who purchased stoves did so to save fuel and also to improve the kitchen environment by reducing smoke levels, kitchen cleaning and the risk of burns. The principal reason for using improved mud stoves was smoke removal and, for ceramic stoves, fuel saving. In northern India (NADA), 40 per cent of the stove users said smoke removal was the main benefit, 30 per cent said it was this combined with fuel savings, 11 per cent gave time savings as the principal reason, and only 4 per cent singled out fuel saving. In Guatemala smoke removal was seen as the principal benefit by 35 per cent, followed by fuel savings at 28 per cent. However, in Indonesia 43 per cent put fuel savings as the main benefit and a further 46 per cent combined fuel savings with greater stove durability as the main advantages.

From the stove producers' point of view, all the mud stove makers (CEMAT, ASTRA,NADA) increased their income by this new activity but not always reliably enough to make it an assured, permanent occupation. In north India (NADA) stove makers had to find other

Table 8.5: Features of Five Leading Improved Stove Programmes

Programme: Country:	Dian-Desa Indonesia	CEMAT Guatemala	ASTRA S. India	NADA N. India	KENGO Kenya
Primary objectives	Fuel	Fuel ID	Fuel	Kitchen WD	Fuel
Stove type Fuel	Ceramic Wood	Mud Wood	Mud Wood	Mud Wood	Metal/ceramic Charcoal
Programme start	1979	1978	1982	1980	1981
Stoves produced* (thousands)	25	2	180	6	180
Programme costs (US$) (per stove produced)	3	3–5	?	3	4–6
Research/development sources	Producer & Lab.	User	Lab. & User	User	Producer & Lab.
Dissemination	Commerc. & NGO	Commerc. & NGO	Govt	Commerc. & NGO	Commerc.
Direct subsidy	No	No	Yes	Yes	No
Average fuel savings (per cent)	62	63	16	25	31
Cost payback time (weeks)	20–70	27	7	2–8	4
% users reporting time savings	29	N/A	92	74	90
% users reporting smoke reduction	N/A	98	90	84	N/A

Notes:
* Produced as a direct result of the input from the promoting organization. With CEMAT/Guatemala, in particular, other organizations have built many more Lorena stoves.
ID = integrated development. WD = women's development.
Commerc. = commercial. Govt = built by government extension workers.

Source: Joseph (1987).[20]

employment when local stove demand had been met – but were usually able to do so. In south India (ASTRA) there was an 80 per cent drop-out rate among stove-making trainees, largely due to lack of continued advice and support. In Guatemala (CEMAT) stoves were made by community groups or micro-enterprises, many of whom were able to turn to other activities when stove demand tailed off. In the Niger programme, which trained nearly 250 stove builders in the first year,[19] stove demand after that time was running at only a tenth of potential capacity. Whereas each trained person could make eight to twelve stoves a day, or a total of 610–790, actual production was only sixty or so stoves a day because of low demand. The small artisanal manufacturers could not afford to keep stocks and had to make each stove to order.

These findings suggest that the next main battle to be won on the stove promotion front lies not so much with meeting the diverse needs of stove users, which are now well recognized, but in underpinning the production and marketing side. As the programme review by the Foundation for Woodstove Dissemination puts it, for the small village or urban stove producer it may be vital to ensure that:

- they can earn more money by making stoves than alternative activities, including for rural artisans the question of whether stove making can be restricted to periods when agricultural labour demands are low
- they receive continued training and technical support from the stove designers or the extension agency promoting the stove, not least to ensure quality control. Good stove designs poorly made can make bad stoves
- credit is available if necessary to buy equipment and materials, or to keep production going if demand falls off temporarily. Initial large orders may be needed to get a collective or artisanal enterprise interested in stove manufacture
- they receive assistance with marketing and business planning if necessary
- they receive assistance in developing other new products if a decline in stove demand threatens to shut down their business.

In short, these points suggest that – as with so many other woodfuel issues – successful stove programmes depend on broader questions than "energy" alone. For stove users, fuel savings are usually only one of the potential advantages of an improved stove. For the producers, stove development must be seen in the wider context of improving the commercial environment and capabilities of artisanal activities.

ENCOURAGING FUEL SWITCHING

Through Prices?

What can a government do to encourage the transition away from woodfuels, assuming that such a shift is thought to be desirable? The most frequently used and seemingly most powerful lever is to control the prices of the main alternatives to woodfuels, namely kerosene and bottled gas. However, this approach has turned out to be more difficult and less effective than one might suppose.

For one thing, there is little hard information on what a given price change might do to alter consumption levels or fuel-switching behaviour, especially for the all-important household sector. In developed countries there is a massive literature on the effects of prices on the relative consumption of competing fuels and of cross-fuel switching, but in the great majority of cases – as with natural gas versus electric cooking and heating, for example – there are no other significant factors at work. Investment costs are broadly the same, supplies are equally reliable and, when comparing these two energy sources, there is complete accessibility for those consumers who are in a position to switch, in the sense that both energy sources are delivered "to the door". There is virtually no equivalent information in developing countries, for lack of adequate data on price and consumption trends.

Second, as we have seen above, the major constraints on fuel switching are: (1) poor access to modern fuels and their unreliable supply; and (2) high equipment costs for modern fuels. Both are aspects of poverty: the second directly and the first more indirectly through the lack of effective demand to force improvements in fuel distribution.

In other words, price controls do little to address the main constraints on fuel switching. This fact was one of the major findings of the study on fuel consumption and prices in nineteen developing countries by Douglas Barnes of the World Bank which was discussed in Chapter 6.[23] Although the findings were weak in the statistical sense there were definite signs that:

1. It was high average income which had the strongest effect on holding down woodfuel prices, since it facilitates substitution into alternative fuels. In other words, it was rising income rather than high woodfuel prices relative to modern fuels that seemed to be the main cause of interfuel substitution up the preference ladder.

2. There was some evidence that sharp increases in kerosene prices pulled up woodfuel prices. More interestingly, low kerosene prices did not reduce the price of woodfuels unless there was at the same time a major shift towards higher or at least more equitable distribution of income. In short, kerosene subsidies, which tend to favour high-income households, had little effect on the almost separate woodfuel markets of

the poor. Once again, it was high or rising income that had the greatest effect on upward fuel switching.

Actual experience of relative price setting to encourage fuel switching tends to bear out these points. One well-known example is the "butanization" policy adopted in the mid 1970s in Senegal. Because of government concern over deforestation, butane (LPG) was heavily subsidized to encourage households to use it instead of charcoal, with the aim of halving charcoal use by 1977.[24] The controlled price of butane became roughly one-fifth the price of charcoal (and one-sixth that of kerosene) on a useful heat basis. What happened was that charcoal use rose slightly but consumption of subsidized butane almost quadrupled from 2,900 tons in 1974 to 11,000 tons in 1981, about half of which had to be imported at considerable cost. The main effect of the large butane subsidy was to persuade kerosene rather than charcoal users to switch into butane.

Or By Improving Access?

Although pricing may have a role to play, there is now a growing consensus that fuel switching has to be encouraged by a more direct attack on the twin barriers which prevent the poor from achieving their aspirations for modern fuels: poor access to, and unreliable supply of, modern fuels and high equipment costs. There are many things that can be done in both regards, but some of them are costly or involve rather indirect approaches.

The most indirect approach, of course, is to do nothing except promote development in the most general sense. As argued earlier, the switch from woodfuels to modern fuels is in the broadest sense driven by urbanization (especially the growth of the largest cities), the modernizing influences this brings, and rising urban incomes. Demand for modern fuels increases and, gradually, the distribution system improves to meet it. In particular, rising demand for modern fuels by industry and the commercial sector – rather than the household sector itself – usually leads this process.

But there are also many more direct approaches to easing the constraints on fuel switching which can be tried. *Access to fuels* and the *reliability of supplies* can be improved in towns or in parts of larger cities where they are now poor by, for example:

- improving facilities for transporting kerosene and bottled gas within cities, especially to poorer districts and to smaller outlying towns; lack of transport is a frequent cause of supply failures and maldistribution of supplies

- improving storage facilities at the refinery or on the docks to smooth out supply bottlenecks
- improving the supply of bottled gas cylinders; chronic shortages of these constitute a major reason why bottled gas is not used by households and small commercial or industrial users who would otherwise turn to it from woodfuels
- providing support to retailers, or encouraging the expansion of retail outlets, especially in disadvantaged areas; possible mechanisms might include financial help to support the maintenance of larger stocks, or the introduction of "sale or return" for retailers by the main distribution organisations.

Costs of modern fuel devices are the second main barrier to upward switching for all but high-income households. These devices must also be appropriate to user needs. Many possible mechanisms can be envisaged for reducing these costs and improving the quality of product, including:

- encouraging local design, manufacture and dissemination of appropriate, efficient and hopefully low-cost kerosene and LPG cookstoves (and kerosene lamps). At present, equipment is often imported, resulting in high consumer prices and foreign exchange costs; imported equipment may also be difficult to obtain or may not satisfy consumer needs for various reasons. If local manufacture could be built up sufficiently, economies of scale in production and marketing could result in substantial unit cost reductions
- subsidizing kerosene stoves and bottled gas stoves and cylinders. A problem here is how to target subsidies to poorer families, rather than potential switchers in the upper-income range. Models exist for targeting fuel subsidies to the poor, such as the low-price kerosene stamp scheme which operates in Sri Lanka and the government-run low-price stores sited in the poorer districts of many Indian cities
- establishing loan or stamp schemes to enable poorer families to spread the often prohibitively high first cost of kerosene and bottled gas equipment
- experimenting with the production and sale of bottled gas in smaller and cheaper cylinders than the usual 12–14 kilogram sizes. This would increase access in two ways, by easing cash flow

problems and difficulties over carrying home large and heavy cylinders.

Conflicts with fuel-saving measures may arise from these objectives. Given the choice of an improved and efficient charcoal stove or an improved kerosene stove, many consumers may opt for the former, depending on investment and running costs and non-economic fuel preferences. But in either case, charcoal and trees would be saved, which is no bad thing.

Improving access to electricity is an enormous topic which has not been discussed here because its rationale is so much broader than the issue of woodfuels and fuel switching. Families and other consumers do not want to adopt electricity only to get out of woodfuels but also to gain all the advantages of better lighting and the many new kinds of energy service which electrification offers, from refrigerators to TV. Similarly, policies to expand electricity distribution and/or to subsidize connection and consumption costs for households are typically made on broad development and welfare grounds rather than to meet fuel-switching objectives.

FOREIGN EXCHANGE CONSTRAINTS?

A final question that has to be asked is whether national economies can afford to leave woodfuels behind by encouraging the transition to modern fuels.

Interestingly, on a global scale the problem is a trivial one in terms of energy consumption and costs, as a crude back-of-the-envelope calculation suggests. According to FAO estimates,[25] probably half the world population, or some 2,500 million people, rely wholly or mainly on woodfuels. Data from many urban household surveys[2] show that if they abandoned woodfuels altogether for kerosene, bottled gas (or natural gas where it is available) their annual consumption of the latter would be roughly 100 kilograms of oil equivalent each, so that extra world oil consumption from this total switch away from woodfuels would be about 250 million tons. This is equivalent to only 9 per cent of world oil consumption; 3 per cent of world use of oil, gas, coal and primary electricity in 1986; and three-quarters of the decline in oil use by the industrial (OECD) countries from 1979 to 1986.[26]

But what about individual and poorer countries? Table 8.6 shows the

Table 8.6: Impacts of Replacing Urban Woodfuels with Kerosene

country/ year	extra kerosene to replace all urban woodfuels ('000 tonnes)	increase in total petroleum due to kerosene replacement (per cent)	decrease in rural and urban woodfuels due to kerosene replacement (per cent)
Low income			
Benin 1983	40	34	26
Burundi 1980	15	46	23
Ethiopia 1982	170	32	26
Ghana 1985	180	26	44
Guinea 1984	110	29	23
Kenya 1985	335	20	29
Malawi 1980	40	29	5
Niger 1980	35	23	20
Nigeria 1980	405	5	12
Rwanda 1978	13	27	8
Sierra Leone 1986	60	34	37
Somalia 1984	75	41	32
Sudan 1980	435	45	32
Tanzania 1981	470	74	29
Togo 1981	40	17	25
Uganda 1982	80	51	15
Zaire 1983	540	91	37
Middle income			
Congo 1985	32	13	50
Ivory Coast 1982	135	14	35
Liberia 1983	50	12	40
Mauritania 1984	13	8	40
Senegal 1981	55	8	27
Zambia 1981	110	16	55
Zimbabwe 1980	65	11	18

Note:
Efficiency assumptions: firewood 10%, charcoal 15%, kerosene 35%.

Source: Byer (1987)[1], based on UNDP/World Bank Country Energy Sector Assessments.

impact on petroleum and woodfuel consumption of substituting with kerosene all urban charcoal and firewood use in twenty-five African countries. The table is based on broad estimates of woodfuel consumption and depends critically on assumptions about the relative efficiencies of woodfuel and kerosene. It should therefore be taken as no more than a rough guide.

Nevertheless, the table does show that the countries do fall into three very distinct groups. The first group comprises all seven of the middle-income countries plus low-income Nigeria, in which a complete switch into kerosene would increase petroleum demand by only 5 to 16 per cent. Such small increases are arguably easily manageable if spread over several years, as would be the case in practice. The effects on woodfuel use are, however, dramatic. With the exception of Nigeria, where the low rise in oil demand is accounted for by exceptionally high oil use in the economy as a whole, for the first group every 1 per cent increase in oil use results in a 4 to 5 per cent reduction in total woodfuel consumption: that is, consumption in rural and urban areas combined.

The second group includes five of the low-income countries (Ethiopia, Ghana, Kenya, Sierra Leone and Togo) where the increase in petroleum demand ranges from 17 to 34 per cent but the fall in woodfuel consumption is proportionately much less. Each 1 per cent rise in oil use translates into only a 2 per cent decline in total woodfuel use.

The third group comprises all the remaining poor countries, where the level of oil use in the economy as a whole is low. Fuel switching in the household sector therefore contributes disproportionately to increased total oil demand. In this group, oil demand rises by roughly 30 per cent to as much as 74 per cent in the case of Tanzania. All told, each percentage rise in oil use results in only a 1 per cent fall in woodfuel consumption.

This comparison emphasizes in a stark way the woodfuel-related dilemma of poorer developing countries, which this book has tried to address. The richer, better developed and more urbanized of the countries have a reasonable chance of buying their way out of woodfuels and following the development path of the North into a modern fuel economy. The rest face a more basic challenge. Importing oil may impose such heavy costs in foreign exchange and increased debt that for the time being it must seem like an impossible option, although one that can never be wholly put aside since it meets basic development aspirations for the majority of energy consumers. But it also does least to help escape from woodfuels. The broad answer, then, must be to accept – for the time being at least – the need for a predominantly woodfuel economy, and try to make the best of it.

As we have tried to demonstrate here, that decision need in no way be seen as "anti-development", an acceptance of "backwardness". If the woodfuel economy and its broader socio-economic context are handled with vision and determination, there are enormous opportunities for easing woodfuel supplies and the stresses they impose on natural resources, while providing multiple benefits within the country for employment, income generation and a more productive and sustainable environment.

Chapter 9

Urban Cases

1: THE FIREWOOD TRADE IN GABORONE, BOTSWANA

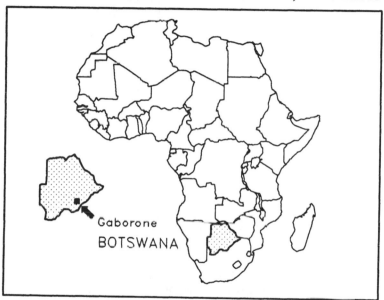

In 1983 a survey was made of all thirty-five traders who collected firewood from the Kweneng district for sale in Gaborone. All were middle-aged men, most had virtually no schooling, and only four lived in the city. Nearly two-thirds (twenty-one) had been in the trade for less than three years and another eight for between four and seven years. About 60 per cent had resorted to firewood selling because they could not find employment: twenty-two had previously worked in the South African mines but had lost their jobs, and ten had been farmers. The second reason for entering the trade was low farm income due to the prolonged drought. Nearly all the thirty-one rural-based traders continued to farm but could not depend on this alone for their livelihoods. Firewood trading was a part-time, second-best choice.

Table 9.1: Income of Firewood Traders in Gaborone

vehicle	typical load (kg)	collection trips per month	gross income per trip (Pula)*	gross income per month (Pula)
1-axle donkey cart	350	2.0	23.5	47.0
2-axle donkey cart	700	1.95	48	93.6
Small trucks	500	4.0	35	140.0†

Notes:
* Including taxes of Pula 1.0 and 2.0 for 1- and 2-axle carts respectively.
† Including costs of about Pula 25 for petrol and labour; no allowance for vehicle maintenance and depreciation.

Of the thirty-five traders, only four operated vehicles, the rest using donkey carts. Incomes from trading depended strongly on the type of vehicle used and the frequency of collection trips. Some details are shown in Table 9.1.

At the time of the survey the minimum wage for unskilled industrial workers was Pula 105 for a twenty-two-day working month. No traders earned more than this (if one allows, say, Pula 10 for additional truck costs not shown in the table) from firewood selling alone. But trading in firewood was a useful extra income to farming or, for the urban-based vehicle owner, to general small-scale business dealing. However, selling firewood was much more profitable than farming, and put the rural traders among the "richer of the poorest".

But despite this, few rural people wanted to take up firewood trading because it was heavy and painful work: most traders complained of back pains from lifting logs. Many of the traders emphasized that they had taken up this work because they had no other means of raising cash incomes.

As for the sources of wood, all the traders said they harvested only dead wood. However, they knew that cutting live wood is prohibited in Kweneng district, and the researchers found definite evidence that live wood was being cut, also that live trees were being burned at the base to kill them for later harvesting as dead wood. About 80 per cent of a sample of local women reported that the firewood trade had contributed to depletion of two species of the most preferred firewood species (*Combretum*) but that there was no shortage of less preferred (*Acacia*) species. They had reduced the frequency of cooked meals, but this was due to drought-related food shortages, not to shortages of fuel.

Source: Kgathi (1984).[1]

2: THE CHARCOAL TRADE IN TANZANIA

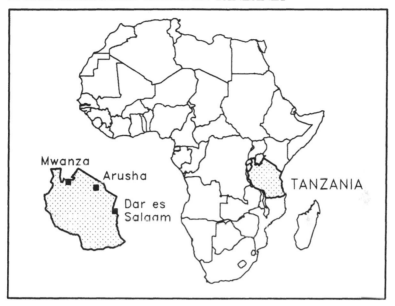

Most charcoal makers in Tanzania use traditional earth or pit kilns and take wood from common land close to main roads. Their methods are rather wasteful since they typically use about 10 tonnes of air-dried wood to make a tonne of charcoal: a weight efficiency of 10 per cent. The range for weight conversions for earth and pit kilns is from 8 to 20 per cent, while more sophisticated low-cost methods such as "beehive" kilns can achieve around 30 per cent.[2] Some charcoalers are full-time operators but most of them turn to making charcoal only in the slack farming seasons to earn some extra income. In the dry season of 1986–7, the ex-kiln price averaged 60–80 shillings for a 35–40-kilogram sack, but could be substantially more or less.

The next stages in the charcoal trade involve a considerable degree of professionalism and integration. Most of the producers sell to registered merchants who buy charcoal at the kiln site, sack it and load it on to trucks for transport to wholesale depots in the main cities. Typically, they use 7-tonne trucks, each carrying about 150 sacks (although with large variations). Some merchants own their own trucks and operate a well-integrated charcoal business, both contracting and financing groups of charcoal makers, and doing their own retail deliveries directly to higher-income households at the other end. Others use hired trucks either as "back haulers" (when trucks carry other goods from the city) or "return transporters" (when trucks travel empty on the outward journey to pick up charcoal).

Transport distances were typically 100–150 kilometres for Dar es Salaam and Dodoma and 200–300 kilometres for Mwanza and Arusha, although some charcoal is carried 400–450 kilometres to Mwanza by rail from the Tabora region. During October 1986, truck hire charges for return haulage were 5,000–7,000 Tanzanian shillings per trip around Dar es Salaam but as much as 18,000 shillings in Mwanza and Arusha, reflecting scarcities and high prices for trucks, spare parts and fuel. For Dar es Salaam transport costs in the dry season of 1986–7 were estimated to be 65–70 shillings[3] and 95 shillings[4] per sack. In the wet season they were almost double this.

The charcoal markets in the cities are highly competitive, with prices reflecting economies of scale and the degree of integration. Wholesale merchants usually deliver sacks to commercial customers such as restaurants and some higher-income households, but their major sales are to retailers. The main costs of the latter include transport within the city to market stalls, and "bulk breaking" into 4-gallon tins ("debes") or small tins usually holding about 1kg, for sale to the mass of consumers who can afford to buy only small amounts on a daily basis. Retailers also have to bear the cost of substantial losses of saleable charcoal – often as much as 15–20 per cent – due to its breaking up in the sack into powder or "fines" during trucking and unloading. These operations plus retailer incomes roughly double the unit price for small-volume sales compared to purchases by the sack.

In the Dar es Salaam markets, prices in 1986–7 ranged from 50 shillings per sack in the Chalinze district to as much as 120 shillings around Chamazi, Mbande and Vikundu. These variations reflect differing transport costs. In the wet season it takes longer to cut wood, dry it and work the kiln, and so prices were generally 20–30 per cent higher.

Several taxes are levied on the charcoal trade by government and local agencies. Charcoal makers are charged 2 shillings a bag ex-kiln by the local government agency, usually a village council or party committee. Transporters are supposed to be charged a tax of 5 shillings per bag at check-points on the main entry points to each of the urban centres. These checks are not difficult to avoid and are sometimes not manned. City councils levy taxes on merchants through licensing of wholesale depots and retailing stalls. Currently the retailing stall licence in Dar es Salaam is 200 shillings a month.

Tables 9.2 and 9.3 summarize these costs and margins, as estimated by the World Bank and SADCC Fuelwood Project studies for the same period of September–October 1986.

Although some of the estimates are quite similar for Dar es Salaam, the differences emphasize how little is known about costs and incomes from such multimillion-dollar trading networks and hence the need for thorough investigations in each case.

Table 9.2: Charcoal Price Structure for Three Tanzanian Cities, Dry and Wet Seasons 1986/7

	dry season			wet season		
	DSM	Dodoma	Mwanza	DSM	Dodoma	Mwanza
	(shillings per charcoal bag)			(shillings per charcoal bag)		
Production						
ex-kiln price	60	70	80	70	80	100
Transport						
average distance (km)	(125)	(125)	(250)	(125)	(125)	(125)
rental	34	12	120	—	—	—
handling	5	5	10	—	—	—
margin	55	37	34	—	—	—
total	94	54	164	174	139	194
Ex-truck price	160	130	250	250	225	300
taxes & levies	7	7	7	5	5	5
retail price (bag)	—	—	—	375	280	450

Notes:
DSM = Dar Es Salaam

Source: UNDP/World Bank (1987).[4]

Table 9.3: Charcoal Price Structure for Dar es Salaam, Dry Season 1986

	shillings/bag	shillings/kg
Ex-kiln price (up to Sh. 120 per sack)	50–80	1.3 – 2.1
Transport, loading/unloading	65–70	1.7 – 1.8
Wholesaler selling price	150	4.0
Retailer selling price		
sack (35–40 kg)	170–180	4.5 – 4.8
debe (4–5 kg)		8 –10
kopo (1–1.1 kg)		8 – 9

Source: McCall (1987).[3]

Sources: McCall (1987),[3] UNDP/World Bank (1987).[4]

3: THE CHARCOAL BUSINESS IN THE SUDAN

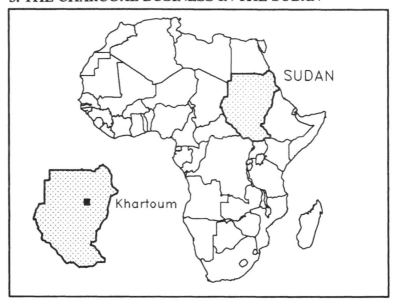

The large urban charcoal markets in the Sudan are extremely sophisticated. In some areas, charcoal production is dominated by networks of entrepreneurs, agents and labourers who finance each other during different seasons as their cash flow situation changes. Huge quantities of charcoal are produced in highly efficient kilns (with a wood-charcoal conversion of about 30 per cent by weight) as much as 500 kilometres from the urban centres.

For Khartoum, about 80 per cent of the supply comes from the Blue Nile and Kassala provinces. Some 85 per cent of this comes from agricultural land clearing schemes and is produced by around twenty-five entrepreneurs (most of whom have been in the business for twenty years or more) who each employ up to three agents. Each agent is responsible for 50 to 100 labourers and oversees production at the charcoal-making sites, including labour recruitment, food and water provisioning, and supply of tools and technical supervision. Most labourers in these agricultural areas are seasonal migrants.

Charcoal production and the way it is financed vary greatly by season. Urban prices are usually highest in the colder February–March period and lowest in the early summer when production is in full swing and the warmer weather reduces demand. Production ceases entirely in the rainy July–November period, largely because transport is too difficult to and from the production sites. Charcoal *storage* is therefore used extensively and there is also a sophisticated system of credit financing among the main sectors of production and marketing. These seasonal features are the following:

September–October	Recruitment of migrant labourers to produce charcoal on land clearance sites; cash advances to labourers as required (up to 20 per cent of total production and marketing costs). Period of highest investment risk by entrepreneurs (e.g. labour may desert after receiving cash advances).
November	Application for charcoaling concessions (concessions granted). Begin burning in small kilns.
February–March	Burning continues. First charcoal sales to take advantage of *highest prices* and to generate working capital for investment in additional labour. Production expands as more labour is recruited.
April–May	Production accelerates to peak rate. Labour bills rise; charcoal sales accelerate to pay labourers. *Lowest prices* due to highest production, lowest demand. Charcoal *storage* in depots begins and production winds down.
June	Production season ends; labour bills settled; excess production is stored.
July–November	Rainy season: production ceases completely. Demand is met by stored charcoal.
September–October	Labour recruitment begins for next production cycle.

The UNDP/World Bank study of the Khartoum market was very thorough and has estimates of many cost items ignored by other studies. The cost structure based on 1983/4 data but with 1986 prices is shown in Table 9.4.

The more an entrepreneur can integrate and control production, transport and distribution, the greater the valued added margin and ability to compete in the market. Losses in one sector which might be incurred in order to gain a particular market share can be offset by profits in another sector. On the other hand, individual wholesalers,

Table 9.4: Price Structure for Urban Charcoal, Khartoum

	price per sack (S£)	per cent of retail price
Retail price	20.0	100
Marketing		17
retailers	2.11	11
wholesalers – 2nd level	1.15	6
wholesalers – 1st level	0.14	1
Price ex-truck	16.61	83
Transport and distribution	7.6	38
(transport to Khartoum)	(3.34)	(17)
(transport to depot)	(1.17)	(6)
(losses and depreciation)	(0.19)	(1)
(handling and packing)	(0.79)	(4)
(guarding at depot)	(1.68)	(8)
(storage tax)	(0.16)	(1)
(depot costs, value added)	(0.27)	(1)
Price ex-kiln (packed in sacks)	9.01	45
(marketing costs, valued added)	(1.18)	(6)
(sacks)	(1.49)	(7)
(fees, royalties, taxes)	(0.80)	(4)
(water 0.59, site restoration 0.19)	(0.78)	(4)
(supervision – agent)	(0.28)	(1)
(supervision – foreman)	(0.45)	(2)
(labourers)	(4.03)	(20)

Source: UNDP/World Bank (1987).[5]

transporters and producers may not be flexible enough to compete in such a sophisticated – and at times, high-risk – system.

Virtually all charcoal is hauled on a *return trip* basis, as a pick-up load on the way back to the city, so transport costs are low. Indeed, charcoal carriage is seen as a means of getting cheaply to Khartoum or Port Sudan to pick up goods there for delivery to other towns and cities in the interior. Even though distances are very large, this means that transport accounts for only 20–25 per cent of the final retail price, a much lower figure than in many other African cities (for example, 45 per cent in Nairobi, Kenya). Key constraints on transport include occasional fuel shortages and poor roads. The former often coincide with shortages of bottled gas, a charcoal competitor: when this happens charcoal prices might rise by 20–25 per cent above the seasonal average.

Retail prices for small-volume sales may be 30 per cent or more above those charged for a sack as shown in Table 9.4; however, this small-volume mark-up is considerably lower than in many African cities.

One of the most remarkable features of the Khartoum charcoal market is the huge seasonal variation in final prices, in real terms. These were discussed briefly in Chapter 6 and shown in Figure 6.4 (p.203). Because there have not been long-term studies of the market, the World Bank study was able only to speculate about the causes of these variations. The main factors they came up with were:

1. *petroleum price increases.* The high peaks in 1978 and 1981–2 could have been due to scarcities of truck fuel and bottled gas which pushed up charcoal transport costs at the same time as charcoal demand increased

2. *good agricultural harvests* in 1981–2 could have contributed to the high prices in those years due to greater demand – and therefore higher wages – for agricultural labourers and itinerant rural workers who make most of the charcoal. Poor harvests during 1979–80 may have had the opposite effect and accounted for the slump in charcoal prices

3. *pressures on capital markets* could have raised interest rates and the costs to entrepreneurs of financing the start-up phase of the charcoaling cycle

4. *high crop prices* could have led to increased rates of land clearance and hence more wood becoming available for charcoal making closer to urban centres, reducing transport costs

5. *collusion among producers and dealers* could have forced up prices for short periods. However, the 1983–4 survey found little evidence of collusion

6. *charcoal exports* to the Middle East have in the past caused scarcities and higher prices but are not currently a feature of the market.

Source: Dewees (1987)[5], El Fakli (1985)[6].

4: HOUSEHOLD FUEL SWITCHING IN DAR ES SALAAM

A survey was made of fuel use and fuel-switching patterns early in 1987 among 230 households in Dar es Salaam, Tanzania. Of the sample, 65 per cent had low incomes, 31 per cent had medium incomes and 4 per cent were in the high-income bracket. Multiple use was common, with woodfuel dominating. All of the bottled gas (LPG) and most of the

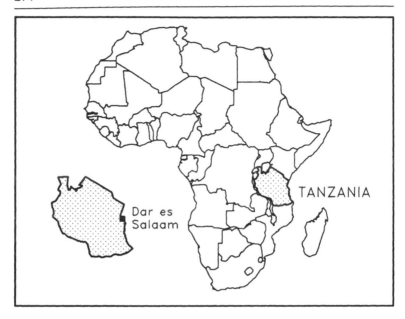

charcoal was used for cooking, but half of the firewood consumption was for other uses (mainly beer brewing) and half of the kerosene use was for lighting. Table 9.5 shows the breakdown of fuels for all of the income groups combined.

As purchased, charcoal was the cheapest fuel per unit of energy and kerosene was the next cheapest. But on a useful energy basis, with (highly uncertain) assumptions about equipment efficiencies, prices fall quite steeply up the income fuel progression, especially from firewood

Table 9.5: Fuel Use in January 1987

	firewood	charcoal	kerosene	LPG	electricity
Households using fuel (%)	11.6	87.0	78.0	7.3	43.0
for cooking (%)	9.1	87.0	43.0	7.3	9.9
Average households:	(kg)	(kg)	(litres)	(kg)	(kWh)
monthly consumption	113.2	98.2	15.5	9.1	41–99
for cooking only	49.6	86.2	7.8	9.1	??
Average households:					
monthly consumption (GJ)	1.8	2.9	0.5	0.4	0.15–0.4
for cooking only (GJ)	0.8	2.6	0.3	0.4	

Source: UNDP/World Bank (1987).[4]

to charcoal. However, only 7 per cent of the sample purchased firewood: the remainer gathered it "free".

Table 9.6: Fuel Prices (Tanzanian Shillings)

	firewood	charcoal	kerosene	LPG	electricity
per unit quantity above	7.9	9.2	14.1	23.1	1.66
per MJ purchased	0.51	0.31	0.40	0.50	0.46
(Assumed efficiency: %)	(10)	(20)	(30)	(50)	(60)
per MJ of useful heat	5.1	1.6	1.3	1.0	0.8

Only 12 per cent of households had switched cooking fuels over the past two years. Of these, 35 per cent had switched "backwards" from kerosene to charcoal, mainly because of kerosene supply difficulties. Another 10 per cent had moved "backwards" into charcoal from bottled gas or electricity even though the latter fuels appear to be considerably cheaper then charcoal. Again, a major reason was supply difficulties. There was much swapping around between firewood and charcoal, and 15 per cent moved up into low-priced electricity. Most notably, no one switched into kerosene or bottled gas, the normal alternatives to woodfuels for the upwardly mobile low- and middle-income family. This switching behaviour is summarized in Table 9.7.

Table 9.7: Percentage of Households Switching Fuels, 1985–7

into:	firewood	charcoal	kerosene	LPG	electricity	
from: firewood		15	—	—	—	"FORWARD SWITCHING"
charcoal	25		—	—	10	
kerosene	—	35		—	—	
LPG	—	5	—		5	
electricity	—	5	—	—		

"BACKWARD SWITCHING"

A great variety of reasons was given for fuel switching, as summarized in Table 9.8. While lower costs for the new fuel was the most common sole reason, convenience of use and supply difficulties were more important reasons all told. Fuel efficiency (that is, the new fuel was "longer lasting") was only a minor consideration.

Table 9.8: Reasons for Switching Fuels

main reason for switch		reason for choosing new fuel (% of switchers)			
Cheaper	38	Cheaper	25	Convenient	17
Greater convenience	28	Cheaper/easy to use	14	Food tastes	5
Supply difficulty	14	Cheaper/convenient	2	Longer lasting	5
No alternative	10	Easy to use/obtain	30	Only fuel available	2

Source: UNDP/World Bank (1987).[4]

5: FOREST TAXES IN MALAWI

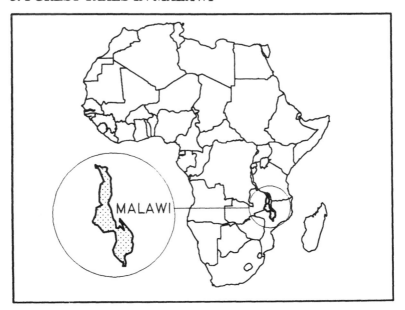

In March 1986 the government raised the stumpage royalty for commercial fuelwood from state forests by 50 per cent, from kwacha 2.7 per stacked cubic metre for indigenous hardwoods and kwacha 2.25 per cubic metre for (plantation) softwoods. Wood which was stacked by the forest department before sale was slightly dearer at kwacha 3.6 per stacked cubic metre for hardwoods and kwacha 3.15 for softwoods.[7] However, these rates are around three to three and a half times less than the costs of kwacha 10–12 per cubic metre for producing wood from forest department plantations in late 1986.[3]

To close this gap, the Department is considering the suggestion by the World Bank Second Wood Energy Project mission that stumpage royalties should be raised by some 15 per cent each year, in real terms, until they reach kwacha 10.7 per stacked cubic metre (kwacha 15.3 per solid cubic metre) in 1996. This is about half the retail price of firewood in Lilongwe, as of 1986, namely kwacha 20 per stacked cubic metre.[7]

The "gradual" tax increase is said to be necessary to allow the Department to build up a large forestry control and revenue collection system and get this "firmly in place".[7] At present the Department's administrative resources are severely stressed by its new responsibilities of revenue collection from and management of 3.7 million hectares of forests on customary land. However, as part of the Second Wood Energy Project, its policing strength would be built up until it had 27 area control units, each of 10 to 15 forest management teams, to control 3,000 hectare apiece. In addition, 17 revenue collection points would be established on major transport routes into Blantyre, Lilongwe and other urban centres, each to be manned by six persons on a 24-hour basis. The additional staff for these policing functions totals around 500 people.

Curiously, while this large team would be trained in the difficult dual role of protecting the forest and "obtaining the cooperation of the [local] people for using the resource on a sustainable basis", only sixty new staff would be taken on under this plan to strengthen the twenty-seven persons who in 1986 formed the Forestry Extension Unit. However, an important feature of the plan is to transfer a portion of the revenues into large subsidies for the establishment of smallholder woodlots, which are estimated to have discounted economic costs of under 30 per cent of those for government plantations (kwacha 258 compared with kwacha 894 per hectare).

The World Bank has considered the effects of these taxes on the fuelwood market and consumers (mostly using estimated data) and concluded that they will not be so great as to cause serious problems. Some key points are shown in Table 9.9.

Half way through the scheme (in 1991) the retail fuelwood price could rise by 18 to 31 per cent in real terms. Since, according to these studies, typical Lilongwe households spend 15 per cent of their income on woodfuels, this would be equivalent to a decline in real incomes of 2.5 to 5 per cent. By the end of the scheme, dealers become unable to absorb all the increased tax (see Item 2 of Table 9.9) and consumer prices could rise by as much as 45 to 75 per cent – unless large,

Table 9.9: Effects of Proposed Increases in Stumpage Royalties on Firewood Prices in Lilongwe, Malawi

	stumpage royalty: kwacha per stacked cubic metre		
	(1983/4)	*(1991)*	*(1996)*
	1.8	*5.4*	*10.7*
1. Basic data (for 22 m³ stacked, or one 14–15 tonne truck):			
purchase of wood	39.6		
cutting & stacking	13.2		
transport (100 km round trip)	105		
loading, unloading, chopping	96.2		
market fees	7		
	261		
sales (kwacha 20 per m³ stacked)	440		
net revenue (kwacha)	179		
"profit" margin (440/261)	69%		
net revenue per day (kwacha)	12.8		
2. Dealers bear all burdens of tax increase:			
net revenue per day (kwacha)	12.8	7.1	–1.2
% price increase to consumers	—	0	0
3. Dealers maintain net revenue per day:			
% price increase to consumers	—	18	45
4. Dealers raise price to maintain profit margin:			
% price increase to consumers	—	31	75

Source: World Bank (1986).[7]

unanticipated changes occur in the market price and cost structure.

In summary, this scheme could have all the faults for which such strategies are often criticized. An effective taxation system requires a large forest police force, with possibly damaging effects on collaborative partnerships with rural people. In practice, the effects in terms of reducing deforestation are at least questionable and could be little or nothing. Consumer prices could rise steeply, but this is not certain since there have been no proper studies of the costs, prices and incentives in the urban markets. If they did rise, consumer costs would have to be held down by vigorous efficiency improvements (as the World Bank study itself acknowledges). But since costs are already quite high, these measures could be initiated straight away, probably with more certain prospects of saving wood resources than the proposed taxation scheme.

Alternatively, high costs would most likely promote structural changes in the market which would make sustainable production more economic anyway. Such changes could perhaps also be initiated directly.[7]

6: SUCCESSFUL FOREST CONTROLS IN RWANDA

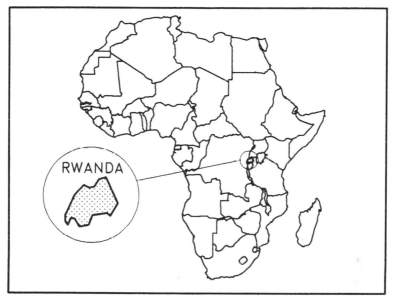

Early in 1987 the government of Rwanda enacted a series of measures to prevent the exploitation of savanna woodlands in the east of the country for charcoal, especially for sale to the capital Kigali. The measures were sufficiently tough to force, in only a few months, a profound switch of charcoal production from the "open access" woodlands of the east (around Bugesera and Kibungo) to small private and communal woodlots in the south and south-west of the country. By October 1987, 85 per cent of Kigali's charcoal was coming from these sources and only 15 per cent from natural woodlands. The biggest change was in the Gikongoro district, which in the same month was supplying 55 per cent of Kigali's charcoal, 95 per cent of it from private or communal woodlots, mostly of eucalypts but also acacias and black wattle (*Acacia mearnsii*).

One enforcement measure was a huge reduction in the number and much stricter control of licences to transport charcoal from the eastern zone. In 1985 some 64 transport permits were issued, mostly for this zone. In the first nine months of 1987 only 16 permits were issued, 14 of them for the major new production area of Gikongoro; and 61 permits were refused. A stiff tax was also imposed on wood from public forests, of 500 Rwanda francs per stacked cubic metre. Official charcoal prices were abandoned in favour of a floating price in order to benefit private producers who were not too far from Kigali. Some communities in the eastern zone also took their own steps to control exploitation of the forests. In the worst-affected area of Bugesera, some communities imposed taxes of 300–1,000 francs on each charcoal truck passing through their territory and the community of Nyamata even closed down its charcoal market by popular agreement.

At the same time, the government introduced a National Forestry Fund to channel the revenues from these measures into the support of more sustainable forestry practices, including reafforestation in the degraded woodlands of the east, seedling nursery improvements, grants or subsidies to support private and community woodlots, and the popularization of forestry techniques.

Another part of the plan was the division of the country into 143 zones based on existing communities and enactment of a law making it obligatory for each zone to produce a forestry management plan for all woodlots of over 2 hectares in size, whether owned by the community, individuals or groups of individuals. Permits to cut these woodlots are required and are issued only on the undertaking that whatever area is cut is replaced by new plantings.

It is too early yet to say whether these profound changes will be lasting. As of October 1987 there were no signs that transporters were trying to break the ban on use of the eastern woodlands by illicit night-time operations and the like. There is also some concern that charcoal making in the new production areas of the south and south-west is much less efficient than it was in the now banned eastern zone because there has been less experience of the techniques involved. This means not only that wood is used more wastefully than it might otherwise be, but also that prices paid to charcoal producers may have to rise if the new sustainable production system is to remain economic. Part of the forestry fund will be used to teach the new producers better charcoal-making techniques.

Source: Matly (1987).[8]

References and Notes

INTRODUCTION

1. FAO, *Fuelwood Supplies in the Developing Countries* (Rome: UN Food and Agriculture Organization, 1983).
2. Anderson, D. and Fishwick, R., *Fuelwood Consumption and Deforestation in African Countries* (Washington, DC: World Bank, 1984).
3. Anderson, D., "Declining Tree Stocks in African Countries", *World Development*, 14/7 (1986): 853–63.
4. Anderson, D. *The Economics of Afforestation: a Case Study in Africa* (Baltimore and London: Johns Hopkins University Press, for the World Bank, 1987).
5. Spears, J., *Deforestation, Fuelwood Consumption, and Forest Conservation in Africa: an Action Program for FY86–88* (Washington, DC: World Bank, 1986). (Restricted.)
6. Nkonoki, S. and Sorensen, B., "A Rural Energy Study in Tanzania: the Case of Bundilya Village", *Natural Resources Forum*, 8 (1984): 51–62.
7. Dewees, P. A., "Commercial Fuelwood and Charcoal Production, Marketing, and Pricing in Kenya". Working Paper II, Phase I Report, *Kenya: Peri-Urban Charcoal/Fuelwood Study* (Washington, DC: World Bank, 1987).
8. Leach, G., *Household Energy in South Asia* (London and New York: Elsevier Applied Science Publishers, 1987).
9. Mazambani, D., "Peri-urban Deforestation in Harare", *Proceedings of the Geographical Association of Zimbabwe*. 14 (1983): 66–81.
10. WRI and IIED, *World Resources 1987* (Washington, DC: World Resources Institute; London: International Institute for Environment and Development; 1987).
11. ETC 1987. A composite reference to the collection of reports of the study *SADCC Energy Development: Fuelwood*. The separate titles, all preceded by *Wood Energy Development:*, are: *Policy Issues*; *A Planning Approach*; *Biomass Assessment*; *The LEAP Model*; *Bibliography of the SADCC Region*; *Country Reports for the Nine SADCC Member States* (Leusden, Holland: ETC Foundation; Luanda, Angola: SADCC Energy Technical and Administrative Unit).
12. ERL, *A Study of Energy Utilisation and Requirements in the Rural Sector of Botswana: Final Report* (London: Environmental Resources Limited, for Overseas Development Administration, 1985).
13. Du Toit, R.F., Campbell, B.M., Haney, R.A. and Dore, D., *Wood Usage and Tree Planting in Zimbabwe's Communal Lands: a Base line Survey of Knowledge, Attitudes and Practices* (Harare: Resource Studies, for the Forest Commission of Zimbabwe and the World Bank, 1984).
14. Gamser, M. *Power from the People: Innovation, User Participation, and Forest Energy Development* (London: Intermediate Technology Publications, 1988).
15. Dewees, P., "The Woodfuel Crisis Reconsidered" (Oxford: Oxford Forestry Institute, 1988) (Draft paper).

16. Tinker, I., "The Real Rural Energy Crisis: Women's Time", *Energy Journal*, 8 (1987):125–46.
17. Cecelski, E., *The Rural Energy Crisis, Women's Work and Family Welfare: Perspectives and Approaches to Action* (Geneva: International Labour Office, 1984).

1. TREES FOR RURAL PEOPLE

1. Foley, G. and Barnard, G., *Farm and Community Forestry*. Earthscan Energy Information Programme Technical Report No. 3 (London: IIED/Earthscan, 1984).
2. FAO, *Tree Growing by Rural People*. Forestry Paper No. 64 (Rome: UN Food and Agriculture Organization, 1986).
3. Richards, P., "Community Environmental Knowledge in African Rural Development", *IDS Bulletin*, 10/2 (1979): 28–36. (Brighton: Institute of Development Studies.)
4. Brokensha, D., Warren, D.M. and Werner, O. (eds), *Indigenous Knowledge Systems and Development* (Lanham: University Press of America, 1980).
5. Chambers, R., *Rural Development: Putting the Last First* (Harlow: Longman, 1983).
6. Richards, P., *Indigenous Agricultural Revolution* (London: Hutchinson, 1985).
7. Stocking, M. and Abel, N., "Ecological and Environmental Indicators for the Rapid Appraisal of Natural Resources", *Agricultural Administration*, 8/6 (1981): 473–84.
   Stocking, M., "Extremely Rapid Soil Survey: an Evaluation from Tanzania", *Soil Survey and Land Evaluation*, 3/2 (1983): 31–6. (Norwich, UK: Geo Books.)
   Bradley, P.N., *Peasants, Soils and Classification: an Investigation into a Vernacular Soil Typology from the Guidimaka of Mauritania*. Research Series No. 14 (Dept of Geography, University of Newcastle upon Tyne, 1983).
   Acres, B.D., "Local Farmers' Experience of Soils Combined with Reconnaissance Soil Survey for Land Use Planning: an Example from Tanzania", *Soil Survey and Land Evaluation*, 4/3 (1984): 77–86. (Norwich, UK: Geo Books.)
   Jungerius, P.D., *Perception and Use of the Physical Environment in Peasant Societies*. Reading Geographical Papers No. 93 (Dept of Geography, University of Reading, 1986).
8. Wisner, B., "Rural Energy and Poverty in Kenya and Lesotho: All Roads Lead to Ruin", *IDS Bulletin*, 18/1 (1987): 23–9. (Brighton: Institute of Development Studies).
9. Richards, P., "Agriculture as a Performance". Paper for *Workshop on Farmers and Agricultural Research: Complementary Methods* (Brighton: Institute of Development Studies, 1987).
10. Chambers, R. and Ghildyal, B.P. "Agricultural Research for Resource-poor Farmers: the Farmer-First-and-Last Model". *IDS Discussion Papers*, DP203 (Brighton: Institute of Development Studies, 1985).
11. Farrington, J. and Martin, A., *Farmer Participatory Research: A review of Concepts and Practices*. Discussion Paper 19 (London: Overseas Development Institute, 1987).
12. Chambers, R. and Jiggins, J. "Agricultural Research for Resource-poor Farmers: a Parsimonious Paradigm". *IDS Discussion Papers*, DP220 (Brighton: Institute of Development Studies, 1986).
13. McCall, M., *Indigenous Knowledge Systems as the Basis for Participation: East African Potentials*. Working Paper No. 36 (Technology and Development Group, University of Twente, Netherlands, 1987).
14. Biggs, S.D. and Clay, E.J., "Sources of Innovation in Agricultural Technology", *World Development*, 914 (1981): 321–36.

15. Swift, J., "Notes on Traditional Knowledge, Modern Knowledge and Rural Development", *IDS Bulletin*, 10/2 (1979): 41–3. (Brighton: Institute of Development Studies.)

16. Rocheleau, D., "Program Strategy Paper for Rural Poverty and Resources: Addressing Landscape Domestication in Marginal Lands Through Free-based Innovations in Policy, Management and Technology" (Nairobi: Ford Foundation, 1986). (Unpublished.)

17. Chambers, R., "Normal Professionalism, New Paradigms and Development". *IDS Discussion Papers*, DP227 (Brighton: Institute of Development Studies, 1986).

18. Shaikh, A.M., Arnould, E., Christophersen, R.H., Tabor, J. and Warshall, P., "Opportunities for Sustained Development: Successful Natural Resources Management in the Sahel". Draft report (Washington, DC: Energy/Development International, 1988).

19. Barrow, E.G.C., "Value of Traditional Knowledge in Present Day Soil Conservation Practice: the Example of the Pokot and Turkana". Paper presented to the *Third National Workshop on Soil and Water Conservation*, September 1986, Nairobi, Kenya.

20. Barrow, E.G.C., "Extension and Learning: Examples from the Pokot and Turkana Pastoralists in Kenya". Paper for *Workshop on Farmers and Agricultural Research: Complementary Methods* (Brighton: Institute of Development Studies, 1987).

21. Mearns, R.J., Kenya trip report, May 1988, International Institute for Environment and Development. (Unpublished.)

22. van Bergen, T., "Joseph Mogaka's Farm: Diversified Biomass Production", *ILEIA Newsletter*, 3/2 (1987): 16–17. (Leusden, Netherlands: Information-centre for Low External-Input Agriculture.)

23. Barrow, E.G.C., Results and findings from a survey on *Ekwar* carried out from November 1986 to July 1987 (unpublished). Lodwar, Turkana District, Kenya: Forestry Department.

24. Haugen, T., "Tree Planting Practices in Dry Areas: Experiences from Konso Development Programme (KDP), Konso, South Ethiopia". Paper for *Workshop on Practical Methods for Community Land Management in the African Drylands* (Hurdalsjoen, Norway: CARE/NORAGRIC, 1987).

25. Tapp, C., "Natural Resources Planning: Responsibilities for the International NGO". Paper for *Workshop on Practical Methods for Community Land Management in the African Drylands* (Hurdalsjoen, Norway: CARE/NORAGRIC, 1987).

26. Tapp, C., Resch, T., Buck, L. and Ntiru, L., *Uganda Village Forestry Project Evaluation* (Kampala, Uganda: CARE-Uganda/Uganda Forest Department, 1986).

27. Fowler, A., Barrow, E., Obara, D. and Odera, P., *An Evaluation of the CARE-International in Kenya Agroforestry and Energy Project* (Nairobi, Kenya: CDP Consultants, 1986).

28. ETC, *Wood Energy Development: a Planning Approach* (Leusden, Netherlands: ETC Foundation; Luanda, Angola: SADCC Energy Technical and Administrative Unit; 1987).

29. Hoskins, M.W., "Community Participation in African Fuelwood Production, Transformation and Utilisation". Paper for *Workshop on Fuelwood and Other Renewable Fuels in Africa* (Paris: Overseas Development Council; Washington, DC: US Agency for International Development; 1979).

30. McGahuey, M., *Assessment of the Acacia Albida Extension Projects in Chad* (Washington, DC: Chemonics, 1985).

31. Swedforest Consulting AB, *Report from the Joint Mission on Village Forestry and Industrial Plantations in Tanzania* (Danderyd, Sweden: Swedforest Consulting AB; Dar es Salaam, Tanzania: Ministry of Natural Resources and Tourism; 1986).

32. Dewees, P. A., "Economic Issues and Farm Forestry in Kenya". Working paper for:

UNDP/World Bank *Kenya Forestry Subsector Review* (Washington, DC: World Bank, 1987).

33. Chambers, R., "Sustainable Livelihoods, Environment and Development: Putting Poor Rural People First". *IDS Discussion Papers*, DP240 (Brighton: Institute of Development Studies, 1988).

34. Charreau, C. and Vidal, P., "Influence de l'Acacia albida sur le sol nutrition minerale et les rendements des mils Pennisetum au Senegal", *Agronomie Tropicale*, 67, (1965).

35. Nkaonja, R.S.W., *Agroforestry: What Is Its Future?* Paper presented at a farewell symposium in honour of Prof. O.T. Edje, 19 July, at Banda College of Agriculture, Malawi (Lilongwe, Malawi: Forestry Department, 1985).

36. Williams, P.J. *Much-Needed Research.* Newsletter PJW-22 (Hanover, NH, USA: Institute of Current World Affairs, 1986).

37. Nyirahabimana, P., "Les arbres fourragers et fruitiers au Rwanda", *Compte-Rendu, Journées d'études forestières du 9 au 12 octobre 1984* (Institut des Sciences Agronomiques du Rwanda, Ruhande).

38. Bognetteau-Verlinden, E., *Study on Impact of Windbreaks in Majjia Valley, Niger* (Wageningen, Netherlands: Wageningen Agricultural University; Niamey, Niger: CARE-Niger; 1980).

39. Edje, O.T., "Agroforestry: Preliminary Results of Interplanting Gmelina with Beans, Maize and Groundnuts", *Luso Journal of Science*, 3/1 (1982): 29–32.

40. Felker, P., *State of the Art: Acacia Albida as a Complementary Permanent Intercrop with Annual Crops* (Washington, DC: US Agency for International Development, 1978).

41. Harrison, P., *The Greening of Africa* (London: IIED-Earthscan; Paladin; 1987).

42. Jensen, O.B., "Rehabilitation in the Sahel: the Majjia Valley Experience: A Way Forward or a Blind Alley?" Paper for *Workshop on Practical Methods for Community Land Management in the African Drylands* (Hurdalsjoen, Norway: CARE/NORA-GRIC, 1987).

43. Kang, B.T., Wilson, G.F. and Lawson, T.L., *Alley Cropping: a Stable Alternative to Shifting Cultivation* (Ibadan, Nigeria: International Institute of Tropical Agriculture, 1985).

44. Nair, P.K.R., *Soil Productivity Aspects of Agroforestry.* Science and Practice of Agroforestry No. 1 (Nairobi, Kenya: International Council for Research in Agroforestry, 1984).

45. Reij, C., "The Agroforestry Project in Burkina Faso: an Analysis of Popular Participation in Soil and Water Conservation". Paper for *Only One Earth Conference on Sustainable Development* (London: International Institute for Environment and Development, 1987).

46. Rorison, K.M. and Dennison, S.E., *Majjia Valley Windbreak Evaluation: Windbreak and Windbreak Harvesting Influences on Crop Production* (Niamey, Niger: CARE-International, 1986).

47. Vonk, R.B., "Farming Systems and Agroforestry Interventions". Paper for *Workshop on Practical Methods for Community Land Management in the African Drylands* (Hurdalsjoen, Norway: CARE/NORAGRIC, 1987).

48. Spears, J., "Review of World Bank Financed Forestry Activity FY1984", in Gregersen, H.M. and Elz, D., *Readings on Social Forestry Projects* (Washington, DC: Economic Development Institute, World Bank, 1986).

49. Dewees, P.A. Personal communication from Oxford: Oxford Forestry Institute, 1988.

50. Refsdal, R., "Small is Beautiful: the Local Markets Should Be Studied". Paper for *Workshop on Practical Methods for Community Land Management in the African Drylands* (Hurdalsjoen, Norway: CARE/NORAGRIC, 1987).

51. van Gelder, B., "The Development and Improvement of Woody Biomass Systems", *ILEIA Newsletter*, 3/2 (1987): 14–15. (Leusden, Netherlands: Information-centre for Low External–Input Agriculture.)
52. Chavangi, N. and Ngugi, A.W., "Innovatory Participation in Programme Design: Tree Planting for Increased Fuelwood Supply for Rural Households in Kenya". Paper for *Workshop on Farmers and Agricultural Research: Complementary Methods* (Brighton: Institute of Development Studies, 1987).

2. FORESTRY FOR LAND MANAGEMENT

1. Arnold, J.E.M., "Economic Considerations in Agroforestry", in Steppler, H.A. and Nair, P.K.R. (eds), *Agroforestry: a Decade of Development* (Nairobi, Kenya: International Council for Research in Agroforestry, 1987).
2. Rocheleau, D., *"Program Strategy Paper for Rural Poverty and Resources: Addressing Landscape Domestication in Marginal Lands Through Tree-based Innovations in Policy, Management and Technology"* (Nairobi: Ford Foundation 1980). (Unpublished.)
3. Budowski, G., "Applicability of Agroforestry Systems", in Gregersen, H.M. and Elz, D., *Readings on Social Forestry Projects* (Washington, DC: Economic Development Institute, World Bank, 1984).
4. Foley, G. and Barnard, G., *Farm and Community Forestry*. Earthscan Energy Information Programme Technical Report No. 3 (London: IIED/Earthscan, 1984).
5. ICRAF, *Agroforestry Research and Development: ICRAF At Work* (Nairobi, Kenya: International Council for Research in Agroforestry, 1987).
6. Nair, P.K.R., *Agroforestry Systems in Major Ecological Zones of the Tropics and Subtropics*. ICRAF Working Paper No. 47 (Nairobi, Kenya: International Council for Research in Agroforestry, 1987).
7. Nair, P.K.R. "Classification of Agroforestry Systems", *Agroforestry Systems*, 3 (1988): 97–128. (Also ICRAF Reprint No.23, Nairobi, Kenya: International Council for Research in Agroforestry.)
8. Hammer, T., *Reforestation and Community Development in the Sudan*. DERAP Publications No. 150 (Bergen, Norway: Chr. Michelsen Institute; and Discussion Paper D-73M, Washington, DC: Resources for the Future, 1982).
Hammer, T., "Sustained Reforestation: Structural and Participatory Prerequisites: Examples from the Sudan's Gum Belt". Paper for *Workshop on Practical Methods for Community Land Management in the African Drylands* (Hurdalsjoen, Norway: CARE/NORAGRIC, 1987).
9. Fernandes, E.C.M., O'Kting'ati, A. and Maghembe, J., "The Chagga Homegardens: a Multistoried Agroforestry System on Mount Kilimanjaro", *Agroforestry Systems*, 2 (1984): 73–86.
10. Raintree, J.B., "Agroforestry Pathways: Land Tenure, Shifting Cultivation and Sustainable Agriculture", *Unasylva*, 36/154 (1986): 2–15.
11. FAO, *Changes in Shifting Cultivation in Africa*. Forestry Paper No. 50 (Rome: UN Food and Agriculture Organization, 1984).
12. Hoekstra, D., *"Leucaena Leucocephela Hedgerows Intercropped with Maize and Beans: an Ex Ante Analysis of a Candidate Agroforestry Land Use System for the Semi-Arid Areas in Machakos District"* (Nairobi, Kenya: International Council for Research in Agroforestry, 1984). (Unpublished.)
13. Gumbo, D., Personal communication, from Harare: ENDA-Zimbabwe, 1988.
14. Rocheleau, D. and Hoek, A. van den, The Application of Ecosystems and Landscape Analysis in Agroforestry Diagnosis and Design: a Case Study from Kathama Sub-location, Machakos District, Kenya. ICRAF Working Paper No. 11

(Nairobi, Kenya: International Council for Research in Agroforestry, 1984).

15. Sepasat, *"The Survey of Economic Plants for Arid and Semi-Arid Tropics: Hedgerow Survey"* (Kew, London: Royal Botanic Gardens, 1986. (Unpublished.)

16. Kuchelmeister, G., *State of Knowledge Report on Tropical and Subtropical Hedgerows* (Eschborn, West Germany: German Agency for Technical Co-operation, 1987).

17. Rocheleau, D.E., "Criteria for Re-appraisal and Re-design: Intra-household and Between-household Aspects of FSRE in Three Kenyan Agroforestry Projects". *ICRAF Working Papers* No. 37 (Nairobi, Kenya: International Council for Research in Agroforestry, 1984).

18. Skutsch, M. McCall, *Why People Don't Plant Trees: the Socioeconomic Impacts of Existing Woodfuel Programmes: Village Case Studies, Tanzania.* Energy in Developing Countries Series, D-73P (Washington, DC: Resources for the Future, 1983).

19. Thomson, J.T., "Public Choice Analysis of Institutional Constraints on Firewood Production Strategies in the West African Sahel", in Russell, C. and Nicholson, N. (eds), *Public Choice and Rural Development.* Research Paper R-21 (Washington, DC: Resources for the Future, 1979).
    Thomson, J.T., *Participation, Local Organisation, Land and Tree Tenure: Future Directions for Sahelian Forestry* (Ouagadougou, Burkina Faso: CILSS/Club du Sahel, 1983).

20. Tapp, C., Resch, T., Buck, L. and Ntiru, L., *Uganda Village Forestry Project Evaluation* (Kampala, Uganda: CARE-Uganda/Uganda Forest Department, 1986).

21. Dewees, P.A., "Economic Issues and Farm Forestry in Kenya". Working paper for: UNDP/World Bank *Kenya Forestry Subsector Review* (Washington, DC: World Bank, 1987).

22. NAS, *Agroforestry in the West African Sahel* (Washington, DC: National Academy of Sciences, 1983).

23. Felker, P., *State of the Art: Acacia Albida as a Complementary Permanent Intercrop with Annual Crops* (Washington, DC: US Agency for International Development, 1978).

24. Poulsen, G., "Trees on Cropland: Preserving an African Heritage", *Ceres,*, No. 104 (1985): 24–8.

25. Conant, F.P., "Thorns Paired, Sharply Recurved: Cultural Controls and Rangeland Quality in East Africa", in Spooner, B. and Mann, H.S. (eds), *Desertification and Development: Dryland Ecology in Social Perspective* (New York: Academic Press, 1982).

26. Pacey, A. and Cullis, A. *Rainwater Harvesting* (London: Intermediate Technology Publications, 1986).

3. CONSTRAINTS ON CHANGE

1. Arnold, J.E.M., "Economic Considerations in Agroforestry", in Steppler, H.A. and Nair, P.K.R. (eds), *Agroforestry: a Decade of Development* (Nairobi, Kenya: International Council for Research in Agroforestry, 1987).

2. Wisner, B. *Power and Need in Africa: Basic Human Needs and Development Policies* (London: Earthscan Publications Ltd, 1988).

3. Collins, J.L. "Labour Scarcity and Ecological Change", in Little, P.D. and Horowitz, M.M. with Nyerges, A.E. (eds), *Lands at Risk in the Third World: Local Level Perspectives.* IDA Monographs in Development Anthropology (Boulder, Colo.: Westview Press, 1987).

4. Wisner, B., "Rural Energy and Poverty in Kenya and Lesotho: All Roads Lead to Ruin", *IDS Bulletin*, 18/1 (1987): 23–9. (Brighton: Institute of Development Studies.)

5. O'Keefe, P., "Cycles of Poverty", *ILEIA Newsletter*, 3/1 (1987): 17. (Leusden, Netherlands: Information-centre for Low External-Input Agriculture.)

6. Livingstone, I., *Rural Development, Employment and Incomes in Kenya.* Report prepared for the ILO's Jobs and Skills Programme for Africa (JASPA) (Addis Ababa: ILO/JASPA, 1981).

7. Horowitz, M.M. and Salem-Murdock, M., "The Political Economy of Desertification in White Nile Province, Sudan", in Little, P.D. and Horowitz, M.M. with Nyerges, A.E. (eds), *Lands at Risk in the Third World: Local Level Perspectives.* IDA Monographs in Development Anthropology (Boulder, Colo.: Westview Press, 1987).

8. Hultin, J., *The Predicament of the Peasants in Conservation-Based Development.* Working Paper No. 8 (Göteborg, Sweden: University of Göteborg, 1986).

9. Adams, M.E., "Community Forestry and Forest Policy in Ethiopia: Some Preliminary Thoughts", *ODI Social Forestry Network Newsletter* No. 3 (London: Overseas Development Institute, 1986).

10. Dewees, P.A., "The Woodfuel Crisis Reconsidered". Draft paper (Oxford: Oxford Forestry Institute, 1988).

11. Hosier, R.H, *The Economics of Agroforestry: Obstacles and Incentives to Ecodevelopment* (Washington State University, Pullman, Wash.: Agroforestry Consortium, 1987).

12. van Gelder, B. and Kerkhof, P., *The Agroforestry Survey in Kakamega District; Final Report.* Working Paper No. 6 (Nairobi, Kenya: Kenya Woodfuel Development Programme/Beijer Institute, 1984).

13. Dewees, P.A., "Economic Issues and Farm Forestry in Kenya". Working paper for: UNDP/World Bank *Kenya Forestry Subsector Review* (Washington, DC: World Bank, 1987).

14. Chambers, R. and Longhurst, R., "Trees, Seasons and the Poor", *IDS Bulletin*, 17/3 (1986): 44–50. (Brighton: Institute of Development Studies.)

15. Chambers, R. and Leach, M., "Trees to Meet Contingencies: Savings and Security for the Rural Poor", *IDS Discussion Papers*, DP228 (Brighton: Institute of Development Studies, 1987).

16. Marilyn Hoskins, quoted in Chambers and Leach (1987), see Note 15.

17. Chavangi, N.A., Rutger, J.E. and Jones, V., *Culture as the Basis for Implementing Self-Sustaining Development Programmes* (Nairobi: Beijer Institute, 1985).

18. Brokensha, D. and Riley, B., "Mbere Knowledge of Their Vegetation and Its Relevance for Development", in Brokensha, D., Warren, D.M. and Werner, O. (eds), *Indigenous Knowledge Systems and Development* (Lanham: University Press of America, 1980).

19. Parkin, D.S., *Palms, Wine and Witness* (San Francisco: Chandler Publishing Company, 1972).

20. Fortmann, L. and Riddell, J., *Trees and Tenure: an Annotated Bibliography for Agroforesters and Others* (Madison, Wis.: Land Tenure Center, University of Wisconsin; Nairobi, Kenya: International Council for Research in Agroforestry; 1985).

21. Brokensha, D. and Riley, B., *Vegetation Changes in Mbere Division, Embu.* Working Paper No. 319 (Nairobi: University of Nairobi, 1977).

22. Wisner, B., Gilgen, H., Antille, N., Sulzer, P. and Steiner, D., "A Matrix-Flow Approach to Rural Domestic Energy: a Kenyan Case Study", in Coclin, C., Smit, B. and Johnston, T. (eds), *Demands on Rural Lands: Planning for Resource Use* (Boulder, Colo.: Westview Press, 1987).

23. Fortmann, L., "Land Tenure, Tree Tenure and the Design of Agroforestry Projects" (Madison, Wis.: Land Tenure Center, University of Wisconsin, 1983). (Unpublished.)

24. Shapera, I., *A Handbook of Tswana Law and Custom* (Oxford: International African Institute, 1955).
25. Allan, W., "Land Holding and Land Usage and the Plateau Tonga of Mazabuka District", in Cotran, E. and Rubin, N. (eds), *Readings in African Law, Vol. I* (London: Frank Cass, 1970).
   Gluckmann, M., *Ideas in Barotse Jurisprudence* (New Haven: Yale University Press, 1965.)
26. Heermans, J.G., "The Guesselbodi Experiment: Bushland Management in Niger". Paper for *Only One Earth Conference on Sustainable Development* (London: International Institute for Environment and Development, 1987).
27. Brokensha, D. and Njeru, E.H.N., *Some Consequences of Land Adjudication in Mbere Division.* Working Paper No. 320 (Nairobi: University of Kenya, 1977).
28. Raintree, J.B., "Agroforestry Pathways: Land Tenure, Shifting Cultivation and Sustainable Agriculture", *Unasylva*, vol. 36, no. 154 (1986): 2–15.
29. Seif el Din, "Agroforestry Practices in the Dry Regions", in Buck, L. (ed.), *Proceedings of the Kenya National Seminar on Agroforestry* (Nairobi, Kenya: International Council for Research in Agroforestry, 1981).
30. Murray, G.F., "The Wood Tree as a Peasant Cash Crop: an Anthropological Strategy for the Domestication of Energy", in Foster, C. and Valdman, A. (eds), *Haiti Today and Tomorrow: an Interdisciplinary Study* (Lanham: University Press of America, 1984).
31. Murray, G.F., "Cash Cropping Agroforestry: an Anthropological Approach to Agricultural Development in Rural Haiti", in *Haiti: Present State and Future Prospects* (Racine, Wis.: Wingspread, 1982).
   Conway, F.J., "Case Study: the Agroforestry Outreach Project in Haiti". Paper for *Only One Earth Conference on Sustainable Development* (London: International Institute for Environment and Development, 1987).
32. Brian, J., "The Uluguru Land Usage Scheme: Success and Failure", *Journal of Developing Areas*, 1980: 175–90.
33. James, R.W., *Tenure and Policy in Tanzania* (Nairobi, Kenya: East African Literature Bureau, 1971).
34. Leakey, L.S.B., *The Southern Kikuyu Before 1903: Volume I* (London: Academic Press, 1977).
35. Penwill, D.J., *Kamba Customary Law Notes Taken in the Machakos District of Kenya Colony* (London: Macmillan, 1951).
36. Kajomulo-Tibaijuka, A., "Factors Influencing the Cultivation of Firewood Trees on Peasant Farms: A Survey on Smallholder Banana-Coffee Farms, Kagera Region, Tanzania", *Rural Development Studies*, No. 18 (Uppsala: Swedish University of Agricultural Sciences, 1985).
37. Thomson, J.T., "Peasant Perceptions of Problems and Possibilities for Local-level Management of Trees in Niger and Upper Volta". Paper for African Studies Association Annual Meeting, 15–18 October 1980, Philadelphia.
   Thomson, J.T., Feeny, D.H. and Oakerson R.J. "Institutional Dynamics: the Evolution and Dissolution of Common Property Resource Management", in National Research Council, *Common Property Resource Management*, (Washington, DC: National Academy Press, 1986).
38. Jackson, J.K., Taylor, G.F. and Condé-Wane, C., *Management of the Natural Forest in the Sahel Region* (Ouagadougou, Burkina Faso: Club du Sahel; Washington, DC: US Agency for International Development, Forestry Support Program; 1983).
   Taylor, G.F. and Soumare, M., "Strategies for Forestry development in the Semi-arid Tropics: Lessons from the Sahel", in Wiersum, K.F. (ed.), *Proceedings of the International Symposium on Strategies and Designs for Reforestation, Afforestation*

and Tree Planting (Wageningen, Netherlands: Wageningen Agricultural University; 1984).

Tapp, C., "Natural Resources Planning: Responsibilities for the International NGO". Paper for *Workshop on Practical Methods for Community Land Management in the African Drylands* (Hurdalsjoen, Norway: CARE/NORAGRIC, 1987).

39. Duncan, P., *Sotho Laws and Customs* (Cape Town: Oxford University Press, 1960).
40. Hammer, T., *Wood for Fuel: Energy Crisis Implying Desertification: the Case of Bara, The Sudan*. MA thesis, (Bergen, Norway: University of Bergen, 1977).
41. Hammer, T., *Reforestation and Community Development in the Sudan*. DERAP Publications No. 150 (Bergen, Norway: Chr. Michelsen Institute, and Discussion Paper D-73M; Washington, DC: Resources for the Future, 1982).
42. Tanner, R., "Land Rights on the Tanganyika Coast", *African Studies*, 19 (1960): 14–25.
43. Williams, P.J.W., *Women's Participation in Forestry Activities in Burkina Faso*. Newsletter PJW-17 (Hanover, NH, USA: Institute of Current World Affairs, 1985a). (Also submitted as a Voluntary Paper for the Ninth World Forestry Congress, Mexico City.)
44. Williams, P.J.W., "Women and Forestry". Special Invited Paper for the Ninth World Forestry Congress, July, Mexico City, 1985b.
45. Swanson, R.A., *Gourmantche Agriculture: Part I – Land Tenure and Field Cultivation* (Integrated Rural Development Project, Eastern ORD, BAEP, Upper Volta: USAID, Contract AID-686-049-78, 1979).
46. Arnold, J.E.M. and Campbell, J.C., *Collective Management of Hill Forests in Nepal: the Community Forestry Project* (Washington, DC: National Academy of Sciences, 1985). Table reproduced in FAO, *Tree Growing by Rural People*. Forestry Paper No. 64 (Rome: UN Food and Agriculture Organization, 1986).
47. Dankelman, I. and Davidson, J., *Women and Environment in the Third World: Alliance for the Future* (London: Earthscan Publications Ltd in association with IUCN, 1988).

4. MEETING THE CONSTRAINTS

1. Blair, H.W. and Olpadwala, P.D., *Forestry in Development Planning: Lessons from the Rural Experience*. Westview Special Studies in Social, Economic and Political Development (Boulder, Colo.: Westview Press, 1988).
2. Chambers, R., "Normal Professionalism, New Paradigms and Development". *IDS Discussion Papers*, DP227 (Brighton: Institute of Development Studies, 1986).
3. ETC, *Wood Energy Development: a Planning Approach* (Leusden, Netherlands: ETC Foundation; Luanda, Angola: SADCC Energy Technical and Administrative Unit, 1987).
4. Gamser, M.S., *Power from the People: Innovation, User Participation and Forest Energy Development* (London: Intermediate Technology Publications, 1988).
5. Leach, G., *Household Energy in South Asia* (London and New York: Elsevier Applied Science Publishers, 1987).
6. Buck, L., Personal communication from Nairobi: CARE-Kenya, 1987.
7. Bunch, R., *Two Ears of Corn: a Guide to People-Centred Agricultural Improvement* (Oklahoma City: World Neighbors, 1982).
8. Ostrom, E., "Issues of Definition and Theory: Some Conclusions and Hypotheses", in National Research Council, 1986. *Common Property Resource Management*. Proceedings of the Conference, 21–26 April 1985 (Washington, DC: National Academy Press, 1986).

9. National Research Council, *Common Property Resource Management.* Proceedings of the Conference, 21–26 April 1985, Washington, DC: National Academy Press, 1986).
10. Hunter, G., *Enlisting the Small Farmer: the Range of Requirements.* Agricultural Administration Unit, Occasional Papers No. 4 (London: Overseas Development Institute, 1982).
11. Alitsi, E., Personal communication from Nairobi: Kenyan Energy Non-Governmental Organizations (KENGO), 1987.
12. Toulmin, C., "Trip Report: Dakar Conference on Environment and Development, 12–13 January 1988"(London: International Institute for Environment and Development, 1988). (Unpublished.)
13. Adams, M.E., "Community Forestry and Forest Policy in Ethiopia: Some Preliminary Thoughts". *ODI Social Forestry Network Newsletter* No. 3 (London: Overseas Development Institute, 1986).
14. Republic of Kenya, Office of the President, *District Focus for Rural Development* (Nairobi, Kenya: Government Printer, 1983).
15. Sindiga, I., and Wegulo, F.N., "Decentralisation and Rural Development in Kenya: a Preliminary Review", *Journal of Eastern African Research and Development*, no. 16 (1986): 134–50.
16. Hoskins, M.W., *Women in Forestry for Local Community Development: a Programming Guide* (Washington, DC: Office of Women in Development, USAID, 1979).
17. Williams. P.J.W., *Taking Chances.* Newsletter PJW-28 (Hanover, NH, USA: Institute of Current World Affairs, 1987).
18. Fowler, A., Barrow, E., Obara, D. and Odera, P., *An Evaluation of the CARE-International in Kenya Agroforestry and Energy Project* (Nairobi, Kenya: CDP Consultants, 1986).
19. World Bank, *Kenya Forestry Subsector Review: Main Report*, Vol. I, Washington, DC: World Bank, 1987).
20. Ilchman, W.F., "Decision Rules and Decision Roles", *The African Review*, no. 2 (1972): 219.
21. Chambers, R., *Rural Development: Putting the Last First* (Harlow: Longman, 1983).
22. Molnar, A., *Review of Social Science Methods for Social Forestry.* Report submitted to the Forestry Department, UN Food and Agriculture Organization (Rome: FAO, 1987).
23. ICRAF, *D&D User's Manual: an Introduction to Agroforestry Diagnosis and Design*, ed. Raintree, J.B. (Nairobi, Kenya: International Council for Research in Agroforestry, 1987).
24. Wisner, B., *Power and Need in Africa: Basic Human Needs and Development Policies* (London: Earthscan Publications Ltd, 1988).
25. Raintree, J.B., "The State of the Art of Agroforestry Diagnosis and Design", *Agroforestry Systems*, no. 5 (1987): 219–50.
26. Conway, G.R. and McCracken, J.A., "Rapid Rural Appraisal and Agroecosystem Analysis", in Altieri, M.A. and Hecht, S.B. (eds), *Agroecology and Small Farm Development* (Florida: CRC Press Inc., forthcoming).
27. Conway, G.R., "Agroecosystem Analysis", *Agricultural Administration*, 20 (1985): 31–55.

5. RURAL CASES

1. ETC, *Wood Energy Development: a Planning Approach* (Leusden, Netherlands: ETC Foundation; Luanda, Angola: SADCC Energy Technical and Administrative Unit, 1987).

2. Brandström, P., "Do We Really Learn from Experience? Reflections on Development Efforts in Sukumaland, Tanzania", in Hjort, A. (ed.), *Land Management and Survival* (Uppsala, Sweden: Scandinavian Institute of African Studies, 1985).

3. Barrow, E., Kabelele, M., Kikula, I. and Brandström, P., "Soil Conservation and Afforestation in Shinyanga Region: Potentials and Constrants". Mission Report to NORAD (Dar es Salaam, Tanzania, 1988).

4. Barrow, E.G.C., "Value of Traditional Knowledge in Present Day Soil Conservation Practice: the Example of the Pokot and Turkana". Paper presented to the *Third National Workshop on Soil and Water Conservation*, Nairobi, Kenya, 16–19 September 1986.
   Barrow, E.G.C., "Extension and Learning: Examples from the Pokot and Turkana Pastoralists in Kenya". Paper for *Workshop on Farmers and Agricultural Research: Complementary Methods*, Brighton: Institute of Development Studies, 26–31 July 1987.
   Barrow, E.G.C., Results and findings from a survey on *Ekwar* carried out from November 1986 to July 1987 (Lodwar, Turkana District, Kenya: Forestry Department, 1987).

5. Kenyan Forestry Department, *Rural Forestry: Trees in Our Life (Miti Maishani Mwetu)*. Newsletter No. 14 (Nairobi, Kenya: Extension and Information Services, RAES, 1988).

6. Mearns, R.J., Kenya trip report, May 1988, International Institute for Environment and Development (Unpublished.)

7. Shaikh, A.M., Arnould, E., Christophersen, R.H., Tabor, J. and Warshall, P., "Opportunities for Sustained Development: Successful Natural Resources Management in the Sahel". Draft report (Washington, DC: Energy/Development International, 1988).

8. Haugen, T., "Tree Planting Practices in Dry Areas: Experiences from Konso Development Programme (KDP), Konso, South Ethiopia". Paper for *Workshop on Practical Methods for Community Land Management in the African Drylands*. (Hurdalsjoen, Norway: CARE/NORAGRIC, 1987).

9. Wilson, K.B., "Research on Trees in the Masvihwa and Surrounding Areas". Report for ENDA-Zimbabwe, 1987. (Unpublished.)

10. Scoones, I., Personal Communication from London: Renewable Resources Assessment Group, Imperial College, 1988.

11. Rocheleau, D.E., "The User Perspective and the Agroforestry Research and Action Agenda", in Gholz, H.L. (ed.), *Agroforestry: Realities, Possibilities and Potentials* (Dordrecht, Netherlands: Martinus Nijhoff Publishers in co-operation with ICRAF, 1987).

12. Singh, R., Helgaker, K. and Holden, S.T., *Improved Fallow Systems* (Kasamu, Zambia: Soil Productivity Research Programme, Misamfu Regional Research Station, 1987).

13. Chidumayo, E.N., "A Shifting Cultivation Land Use System under Population Pressure in Zambia", *Agroforestry Systems*, 5 (1987): 15–25.

14. Chambers, R. and Longhurst, R. "Trees, Seasons and the Poor", *IDS Bulletin*, 17/3 (1986): 44–50 (Brighton: Institute of Development Studies).

15. Jama, B., "Learning from the Farmer: What Is the Role of Agricultural Research in Kenya?" Paper for *Workshop on Farmers and Agricultural Research: Complementary Methods* (Brighton: Institute of Development Studies, 1987).

16. Chavangi, N.A., Rutger, J.E. and Jones, V., *Culture as the Basis for Implementing Self-Sustaining Development Programmes* (Nairobi: Beijer Institute, 1985).

17. Farrington, J. and Martin, A., *Farmer Participatory Research: a Review of Concepts and Practices*. Discussion Paper 19 (London: Overseas Development Institute, 1987).

18. Chavangi, N.A., *Cultural Aspects of Fuelwood Procurement in Kakamega District.* Working Paper No. 4 (Nairobi: Kenya Woodfuel Development Programme/Beijer Institute, 1984).
19. Bradley, P.N. "Methodology for Woodfuel Development Planning in the Kenyan Highlands", *Journal of Biogeography,* 15 (1988): 157–64.
20. Ngugi, A.W., "Cultural Aspects of Fuelwood Shortage in the Kenyan Highlands", *Journal of Biogeography,* 15 (1988): 165–70.
21. Reij, C., "The Agroforestry Project in Burkina Faso: an Analysis of Popular Participation in Soil and Water Conservation" Paper for *Only One Earth Conference on Sustainable Development* (London: International Institute for Environment and Development, 1987).
22. Kramer, J.M. "Sustainable Resource Management". Overview paper for *Only One Earth Conference on Sustainable Development* (London: International Institute for Environment and Development, 1987).
23. Pacey, A. and Cullis, A., *Rainwater Harvesting* (London: Intermediate Technology Publications, 1986).
24. Twose, N., *Why the Poor Suffer Most: Drought and the Sahel* (Oxford: OXFAM, 1984).
25. Wright, P. and Bonkoungou, E.G., "Soil and Water Conservation as a Starting Point for Rural Forestry: the OXFAM Project in Ouahigouya, Burkina Faso", *Rural Africana,* nos 23–4 (1984–5): 79–86.
26. Gamser, M.S., "Letting the Piper Call the Tune: Experimenting with Different Forestry Extension Methods in the Northern Sudan", *ODI Social Forestry Network Papers* 4a (London: Overseas Development Institute, 1987).
    Gamser, M.S., *Power from the People: Innovation, User Participation and Forest Energy Development* (London: Intermediate Technology Publications, 1988).
27. Bognetteau-Verlinden, E., *Study on Impact of Windbreaks in Majjia Valley, Niger* (Wageningen, Netherlands: Wageningen Agricultural University; Niamey, Niger: CARE-Niger; 1980).
28. Delehauty, J., Hoskins, M. and Thomson, J.T., *Majjia Valley Study: Sociological Report* (Niamey, Niger: CARE-Niger, 1985).
29. Dennison, S., *Majjia Valley Evaluation Study: Updates Sept. 1984, Dec. 1984, Aug. 1985* (Niamey, Niger: CARE-Niger, 1985).
30. Williams, P.J.W., *(No Longer) Blowin' in the Wind.* Newsletter PJW-15 (Hanover, NH, USA: Institute of Current World Affairs, 1985).
31. Gregersen, H.M., Draper, S. and Elz, D. (eds), *People and Trees: Social Forestry Contributions to Development* (Washington, DC: World Bank, 1986).
32. Harrison, P., *The Greening of Africa* (London: IIED-Earthscan; Paladin; 1987).
33. Jensen, O.B. "Rehabilitation in the Sahel: the Majjia Valley Experience: a Way Forward or Blind Alley?" Paper for *Workshop on Practical Methods for Community Land Management in the African Drylands* (Hurdalsjoen, Norway: CARE/NORA-GRIC, 1987).
34. Östberg, W., *The Kondoa Transformation: Coming to Grips with Soil Erosion in Central Tanzania.* Research Report No. 76 (Uppsala: Scandinavian Institute of African Studies, 1986).
35. McCall-Skutsch, M. and McCall, M., Personal communication: Tanzania case study for ETC 1987, from Enschede, Netherlands: Twente University of Science and Technology, 1986.
36. Nshubemuki, L. and Mugasha, A.G., "The Modifications to Traditional Shifting Cultivation Brought about by the Forest Development Project in the HADO Area, Kondoa, Tanzania", in FAO, *Changes in Shifting Cultivation in Africa: Seven Case Studies.* FAO Forestry Paper 50/1: 141–162 (Rome: FAO, Forest Department, 1985).

37. Mgeni, A.S.M., "Soil conservation in Kondoa District, Tanzania", *Land Use Policy* (July 1985): 205–9.
38. McCall-Skutsch, M., "Forestry by the People for the People: Some Major Problems in Tanzania's Village Afforestation Programme", *International Tree Crops Journal*, vol. 3, nos 2–3 (1985): 147–70.
39. van Gelder, A., Hosier, R. and van der Donk, W., "Fuelwood Production in Developing Countries: Towards an Appropriate Forest Technology", *Proceedings of the Indian Academy of Sciences, Engineering* (New Delhi: Indian Academy of Sciences, 1983).
40. van Gelder, B., Enyola, M. and Mung'ala, P., *Traditional Agroforestry Practices on Farms in Kakamega District*. Kenya Woodfuel Development Programme Working Paper (Nairobi: Beijer Institute, 1985).
41. Hosier, R.H., *The Economics of Agroforestry: Obstacles and Incentives to Ecodevelopment* (Washington State University, Pullman, Wash.: Agroforestry Consortium, 1987).
42. Hagen, R., Diakite, S., Kante, S., Sissako, M. and Dennison, S., *Koro Village Agroforestry Project: an Evaluation Report* (Bamako, Mali: CARE-Mali/Direction Nationale des Eaux et Forêts, 1986).
43. Laumark, S., *CARE-Mali Village Agroforestry Project Annual Report July 1986–June 1987* (Bamako, Mali: CARE—Mali, 1987).
44. Vonk, R.B. *Report on a Methodology and Technology Generating Exercise* (Wageningen, Netherlands: Wageningen Agriculture University, 1983).
45. Hoek, A. van den, *Landscape Planning and Design of Watersheds in the Kathama Agroforestry Project, Kenya*. MSc. thesis (Wageningen, Netherlands: Wageningen Agricultural University, Dept of Landscape Architecture and Planning, 1983).
46. Rocheleau, D., and Hoek, A. van den, *The Application of Ecosystems and Landscape Analysis in Agroforestry Diagnosis and Design: a Case Study from Kathama Sublocation, Machakos District, Kenya*. ICRAF Working Paper No. 11 (Nairobi, Kenya: International Council for Research in Agroforestry, 1984).
47. Wijngaarden, J. van, *Agricultural Self-Help Groups and their Potential Role in Agroforestry* (Wageningen, Netherlands: Wageningen Agricultural University, 1983).
48. ICRAF, *D & D User's Manual: an Introduction to Agroforestry Diagnosis and Design* ed. Raintree, J.B. (Nairobi, Kenya: International Council for Research in Agroforestry, 1987).
49. Skutsch, M. *Why People Don't Plant Trees: the Socioeconomic Impacts of Existing Woodfuel Programmes: Village Case Studies, Tanzania*. Energy in Developing Countries Series, D-73P (Washington, DC: Resources for the Future, 1983).
50. Fraiture, A. de and Hijweege, W.L., *De Konditionering van een "social forestry" programma door machts- en invloedrelties* (Wageningen, Netherlands: Vakgroep Boshuishoudkunde/ Vakgroep Niet-Westers Recht, 1986).
51. USAID, *Short and Long Term Impact of USAID/Haiti's Agroforestry Outreach Project Limited by Lower-than-Expected Tree Survival and Growth Rates and Deficient Research Techniques*. Audit Report No. 1-521-84-8 (Washington, DC: US Agency for International Development, 1984).
52. Conway, F.J. "Case Study: the Agroforestry Outreach Project in Haiti". Paper for *Only One Earth Conference on Sustainable Development* (London: International Institute for Environment and Development, 1987).
53. Kramer, J.M., "Sustainable Resource Management". Overview paper for *Only One Earth Conference on Sustainable Development* (London: International Institute for Environment and Development, 1987).
54. Heermans, J.G., "The Guesselbodi Experiment: Bushland Management in Niger". Paper for *Only One Earth Conference on Sustainable Development* (London:

International Institute for Environment and Development, 1987).

55. Falconer, J., "Forestry Extension: a Review of the Key Issues", *ODI Social Forestry Network Papers* No. 4e (London: Overseas Development Institute, 1987).

56. Taylor, G.F. and Soumare, M., "Strategies for Forestry Development in the Semi-arid Tropics: Lessons from the Sahel", in Wiersum, K.F. (ed.), *Proceedings of the International Symposium on Strategies and Designs for Reforestation, Afforestation and Tree Planting* (Wageningen, Netherlands: Wageningen Agricultural University, 1984).

Taylor, G.F. and Soumare, M., "Strategies for Forestry Development in the West African Sahel: an Overview", *Rural Africana*, nos 23–4 (1984–5): 5–19.

57. Thomson, J.T., *Guesselbodi Forest: Alternative Frameworks for Sustained-Yield Management*. Forestry and Land Use Planning Project (Naimey, Niger: Direction Nationale des Eaux et Forêts/ US Agency for International Development, 1981).

58. Orr, B., *Refugee Forestry in Somalia: the Step Plan Generates Community Involvement*. Staff paper Series No. 22 (Madison, Wis.: InterChurch Response, 1985).

59. van Gelder, B., "Agroforestry Components in Existing Netherlands Assisted Projects in Kenya" (Netherlands: ETC Foundation, 1987). (Unpublished proposal.)

60. Ethiopian Red Cross Society, *Rapid Rural Appraisal: a Closer Look at Rural Life in Wollo* (Addis Ababa: Ethiopian Red Cross Society; London: International Institute for Environment and Development; 1988).

6. PAYING THE PRICE

1. WRI and IIED, *World Resources 1987* (Washington, DC: World Resources Institute; London: International Institute for Environment and Development; 1987).

2. Hardoy, J.E. and Satterthwaite, D., "Urban Change in the Third World: Are Recent Trends a Useful Pointer to the Urban Future?" in Payne, G. and Cadman, D. (eds), *Future Cities* (London: Methuen, 1988).

3. World Bank, *Accelerated Development in Sub-Saharan Africa* (Washington, DC: World Bank, 1981).

4. Typical low-technology charcoal kilns can convert air-dried wood to charcoal with an efficiency (weight for weight) of around 15 per cent. Since charcoal has twice the energy content of wood (30 MJ/kg compared with 15 MJ/kg for firewood) the efficiency of energy conversion is about 30 per cent. Charcoal cookstoves are roughly 25 per cent efficient in producing heat "under the pot", so the overall energy efficiency is about 7.5 per cent. Typical wood cookstoves have an efficiency of about twice this, or 13–15 per cent. However, different assumptions about efficiencies for cookstoves and charcoal kilns (in which weight conversion can range from 8 to 30 per cent) can either widen this gap or close it completely.

5. ETC 1987. A composite reference to the collection of reports of the study *SADCC Energy Development: Fuelwood*. The separate titles, all preceded by *Wood Energy Development:* are: *Policy Issues; A Planning Approach; Biomass Assessment; the LEAP Model; Bibliography of the SADCC Region; Country Reports for the Nine SADCC Member States* (Leusden, Holland: ETC Foundation; Luanda, Angola: SADCC Energy Technical and Administrative Unit; 1987).

6. IPC, *Solid Fuels in Malawi: Options and Constraints for Charcoal and Coal* (Frankfurt: Interdisziplinäre Projekt Consult, 1988).

7. van Ginneken, W. (ed.), *Trends in Employment and Labour Incomes* (Geneva: International Labour Office, 1988).

8. UNDP/World Bank, *Proceedings of the ESMAP Eastern and Southern Africa*

Household Energy Seminar (Washington, DC: World Bank, 1988). Zimbabwe data: paper by Yemi Katerere, "Issues in Household Energy Strategy Formulation: the Zimbabwe Experience"; Addis Ababa data: discussion comment from Ken Newcombe, World Bank.

9. In Zambia, charcoal sales in the cities of the Central and Copperbelt regions were estimated to be US$24 million in 1983 (Chidumayo, E. N. and Chidumayo, S.B.M., *The Status and Impact of Woodfuel in Urban Zambia* (Lusaka: Government Printer, 1984). In Zimbabwe, all woodfuel sales amounted to Z$90-100 million (US$150 million) in the early 1980s: see Katerere, in Note 8.

10. Foley, G., "Discussion Paper on Demand Management", *Proceedings of the ESMAP Eastern and Southern Africa Household Energy Seminar* (Washington, DC: World Bank, 1988).

11. UN FAO, UNEP, World Resources Institute, World Bank, *The Tropical Forestry Action Plan* (Washington, DC, 1987).

12. Barnes, D., "Understanding Fuelwood Prices in Developing Nations" (Washington, DC: World Bank, Agriculture & Rural Development Department, 1986). (Unpublished.)

13. FAO, *Forest Product Prices 1963–82* (Rome: UN Food and Agriculture Organization, 1983).

14. FAO, "Wood Based Energy and Substitution among Fuels in Africa: Model Framework and Basic Data from Eleven Country Reports" (Rome: UN Food and Agriculture Organization, Forestry Department, Statistics and Economic Division, 1987). (Unpublished.)

15. FAO 1987. Phillip Wardle, Personal communication from Rome: FAO Forestry Department, Statistics and Economics Division, 1987.

16. IMF, *International Financial Statistics* (Washington, DC: International Monetary Fund, 1987).

17. Dewees, P.A., "Consultant's Report on Charcoal and Gum Arabic Markets and Market Dynamics in Sudan". Prepared for: UNDP/World Bank Energy Sector Management Assistance Program, Washington, DC, 1987. See also: El Fakli, G., "Charcoal Marketing and Production Economics in Blue Nile". (Khartoum: Energy Research Council, 1985).

18. Leach, G., *Household Energy in South Asia* (London and New York: Elsevier Applied Science Publishers, 1987).

19. Bowonder, B., *et al.*, *Deforestation and Fuelwood Use in Urban Centres* (Hyderabad, India: Administrative Staff College of India, 1985).

20. Dewees, P.A., "Woodfuel Production Models, Stumpage Rates, and Cost Recovery Issues". Working Paper VII, Phase I Report *Kenya: Peri-Urban Charcoal/Fuelwood Study* (Washington, DC: World Bank, 1987).

21. Bertrand A., "Marketing Networks for Forest Fuels to Supply Urban Centres in the Sahel", *Rural Africana*, nos 23–4 (1984–5) 33–47.

22. UNDP/World Bank, *Senegal: Issues and Options in the Energy Sector* (Washington, DC: World Bank, 1983).

23. Dewees, P.A., "Commercial Fuelwood and Charcoal Production, Marketing, and Pricing in Kenya". Working Paper II, Phase I Report, *Kenya: Peri-Urban Charcoal/Fuelwood Study* (Washington, DC: World Bank, 1987). See also: Note 20 and UNDP/World Bank, *Kenya Urban Woodfuel Development Program* (Washington, DC: World Bank, 1987).

24. Matly, M., Personal communication from Paris: Société, Énergie, Environment et Développement (SEED), 1987.

25. UNDP/World Bank, *Mozambique: Issues and Options in the Energy Sector* (Washington, DC: World Bank, 1987).

26. UNDP/World Bank, *Tanzania: Urban Woodfuels Supply Study, Vol. I, Interim*

Report of Phase One (July 1987) (Washington, DC: World Bank, 1987).
27. UNDP/World Bank, *Burkina Faso: Issues and Options in the Energy Sector* (Washington, DC: World Bank, 1986).

7. TREES FOR THE CITIES

1. Taylor, G.F. and Soumare, M., "Strategies for Forestry Development in the West African Sahel: an Overview", *Rural Africana*, 23–4 (1984–85): 5–19.
2. Openshaw, K., Personal communication from Washington, DC: World Bank, Household Energy Unit, 1988.
3. UNDP/World Bank, *Sudan: Wood Energy/Forestry Project: Main report and Annexes (August 1987)* (Washington, DC: World Bank, 1987).
4. Anon, "Amazon Forest Burning Is Beyond Control", London: *The Guardian*, 1 September 1988.
5. Floor, W., "Household Energy in West Africa: Issues and Options" (Washington, DC: World Bank, Household Energy Unit). (Unpublished.)
6. FAO, *Tanzania Wood Supply Demand Survey* (Dar es Salaam: Ministry of Natural Resources and Tourism, 1971).
7. ETC 1987. A composite reference to the collection of reports of the study *SADCC Energy Development: Fuelwood.* The separate titles, all preceded by *Wood Energy Development:* and: *Policy Issues*; *A Planning Approach*; *Biomass Assessment*; *the LEAP model*; *Bibliography of the SADCC Region*; *Country Reports for the Nine SADCC Member States* (Leusden, Holland: ETC Foundation; Luanda, Angola: SADCC Energy Technical and Administrative Unit; 1987).
8. IPC, *Solid Fuels in Malawi: Options and Constraints for Charcoal and Coal* (Frankfurt: Interdisziplinäre Projekt Consult, 1987).
9. Dewees, P;, Personal communication from Oxford, England: Oxford Forestry Institute, 1987.
10. Jackson, J.K., Taylor, G.F. and Condé-Wane, C., *Management of the Natural Forest in the Sahel Region.* Paper to Club du Sahel, OECD/CILSS (Washington, DC: US Agency for International Development, Forestry Support Program, 1983).
11. Onochie, C.F.A., "An Experiment on Controlled Burning in the Sudan Zone", *Proceedings 1st Nigerian Forestry Conference, Kaduna*, 1964.
12. Thomson, J.T., "Participation, Local Organisation, Land and Tree Tenure: Future Directions for Sahelian Forestry". Paper for Club du Sahel/CILSS, Paris, 1983. (Unpublished).
13. Anon, *Exemple de l'impact d'un projet concernant l'aménagement d'une forêt classée sur les populations: cas du projet d'aménagement de la Forêt de Tobor (Basse Casamance)* (Dakar, Senegal: Secrétariat d'Etat aux Eaux et Forêts, 1981).
14. UNDP/World Bank, *Ethiopia: Issues and Options in the Energy Sector* (Washington, DC: World Bank, 1984).
15. Dubré, Y.C., *Coût des Plantations Energétiques – pays du Sahel: revue de la littérature* (Ottowa: Canadian International Development Agency, Direction des Ressources Naturelles, Secteur Forestiere, 1984).
16. UNDP/World Bank, *Kenya Urban Woodfuel Development Program* (Washington, DC: World Bank, 1987).
17. UNDP/World Bank, *Tanzania: Urban Woodfuels Supply Study*, Vol. I, Interim Report of Phase One (July 1987), (Washington, DC: World Bank).

8. FUEL SWITCHING AND SAVING

1. Byer, T., "Review of Household Energy Issues in Africa", Washington, DC: World Bank (1987). (Unpublished.)
2. Leach, G. and Gowen M., *Household Energy Handbook* (Washington, DC: World Bank, 1987).
3. Newcombe, K., Personal communication from Washington, DC: World Bank, 1988.
4. Floor W., "Household Energy in West Africa: Issues and Options" (Washington, DC: World Bank, 1987). (Unpublished.)
5. Leach, G., *Household Energy in South Asia* (London and New York: Elsevier Applied Science, 1987).
6. Natarajan, I., *Domestic Fuel Survey with Special Reference to Kerosene* (New Delhi: National Council of Applied Economic Research, 1985).
7. GOP, *Housing Census of Pakistan 1980: Summary Results* (Population Census Organization, Government of Pakistan, 1982).
8. O'Keefe, P., Raskin, P. and Bernow, S., *Energy and Development in Kenya* (Stockholm: Beijer Institute; Uppsala: Scandinavian Institute of African Studies; 1984).
9. Sharma, R.K. and Bhatia, R.K., "Energy Needs and Low Income Groups in India, in Kumar, M.S. (ed.), *Energy Pricing Policies in Developing Countries* (Geneva: International Labour office, 1987).
10. Nair, K.N. and Krishnayya, J.G., *Energy Consumption by Income Groups in Urban Areas of India* (Geneva: International Labour office, 1985).
11. ETC 1987. A composite reference to the collection of reports of the study *SADCC Energy Development: Fuelwood.* The separate titles, all preceded by *Wood Energy Development:* are: *Policy Issues*; *A Planning Approach*; *Biomass Assessment*; *the LEAP Model*; *Bibliography of the SADCC Region*; *Country Reports for the Nine SADCC Member States* (Leusden, Holland: ETC Foundation; Luanda, Angola: SADCC Energy Technical and Administrative Unit; 1987).
12. FAO, "Wood Based Energy and Substitution among Fuels in Africa" (Rome: UN Food & Agriculture Organization, Forestry Department, 1987). (Unpublished.)
13. Alitsi, E., Personal communication from Nairobi: Kenya Energy Non-governmental Organization (KENGO), 1987.
14. UNDP/World Bank, *Botswana: Issues and Options in the Energy Sector* (Washington, DC: World Bank, 1984).
15. UNDP/World Bank, *Tanzania Urban Woodfuel Supply Study, Volume 1* (Washington, DC: World Bank, Household Energy Unit, 1987).
16. Foley, G. and Moss, P., *Improved Cooking Stoves in Developing Countries* (London: Earthscan, 1983, revised 1985).
17. Smith, K.R., "Biomass Combustion and Indoor Air Pollution: the Bright and Dark Sides of Small Is Beautiful", in *Proceedings of the Conference on Biomass Energy Management in Rural Areas* (Paris: UNESCO; Hyderabad, India: Administrative Staff College of India, 1983).
18. Mengjie, W., "China: the Spread of Firewood and Coal-efficient Stoves in China and Comparison and Appraisal of Firewood-saving Stoves", in *Stoves for People: Proceedings of International Workshop, Guatemala, October 1987* (Utrecht: Foundation for Woodstove Dissemination, 1987).
19. UNDP/World Bank, *Niger Improved Stoves Project: Mid-term Progress Report* (Washington, DC: World Bank, 1986).
20. Joseph, S., "An Appraisal of the Impact of Improved Wood Stove Programmes: Synthesis of Experience", in *Stoves for People: Proceedings of International*

Workshop, Guatemala, October 1987 (Utrecht: Foundation for Woodstove Dissemination, 1987).

21. Foley, G., Moss, P. and Timberlake, L., *Stoves and Trees* (London: Earthscan, 1984).
22. UNDP/World Bank, *Test Results on Charcoal Stoves from Developing Countries* (Washington, DC: World Bank, 1986).
23. Barnes, D., "Understanding Fuelwood Prices in Developing Nations" (Washington, DC: World Bank, Agriculture & Rural Development Department, 1986). (Unpublished.)
24. UNDP/World Bank, *Senegal: Issues and Options in the Energy Sector* (Washington, DC: World Bank, 1983).
25. FAO, *Fuelwood Supplies in the Developing Countries* (Rome: UN Food and Agriculture Organization, 1983).
26. British Petroleum, *Statistical Review of World Energy* (London: British Petroleum Company, 1987.)

9. URBAN CASES

1. Kgathi, D.L., "Firewood Trade Between Botswana's Rural Kweneng and Urban Gaborone: Employment Creation and Deforestation", *Annual Journal, Forestry Association of Botswana*, 1984.
2. Foley, G., *Charcoal Making in Developing Countries* (London: Earthscan, 1986).
3. McCall, M., Personal communication. Data collected during *SADCC Energy Development: Fuelwood* study (see Introduction, Note 11).
4. UNDP/World Bank, *Tanzania: Urban Woodfuels Supply Study, Vol. I, Interim Report of Phase One (July 1987)* (Washington, DC: World Bank, 1987).
5. Dewees, P.A., "Consultant's Report on Charcoal and Gum Arabic Markets and Market Dynamics in Sudan". Prepared for: UNDP/World Bank Energy Sector Management Assistance Program, Washington, DC, 1987.
6. El Fakli, G., "Charcoal Marketing and Production Economics in Blue Nile". (Khartoum: Energy Research Council, 1985).
7. World Bank, *Malawi: Staff Appraisal Report, Second Wood Energy Project* (Washington, DC: World Bank, Eastern Africa Projects Department, 1986).
8. Matly, M., *Secteur charbonnier: prix du bois sur pied et fiscalité du charbon.* Background paper for UNDP/World Bank Rwanda study, October 1987 (Washington, DC: World Bank, 1987).

Index

Printed and bound by CPI Group (UK) Ltd, Croydon, CR0 4YY

23/10/2024

01777674-0013